T0214525

Moscow Lectures

Volume 1

More information about this series at http://www.springer.com/series/15875

Yuri I. Manin

Introduction to
the Theory of Schemes

NATIONAL RESEARCH
UNIVERSITY

Skoltech
Skolkovo Institute of Science and Technology

Yuri I. Manin
Max Planck Institute for Mathematics
Bonn, Germany

Translated from the Russian and edited by Dimitry Leites.
English edited by Andrei Iacob.

ISSN 2522-0314   ISSN 2522-0322   (electronic)
Moscow Lectures
ISBN 978-3-030-08962-7   ISBN 978-3-319-74316-5   (eBook)
https://doi.org/10.1007/978-3-319-74316-5

Mathematics Subject Classification (2010): 14-01

Cover illustration: https://www.istockphoto.com/de/foto/panorama-der-stadt-moskau-gm490080014-
75024685, with kind permission

Printed on acid-free paper

This Springer imprint is published by the registered company Springer International Publishing AG part
of Springer Nature.
The registered company address is: Gewerbestrasse 11, 6330 Cham, Switzerland

Preface to the Book Series *Moscow Lectures*

You hold a volume in a textbook series of Springer Nature dedicated to the Moscow mathematical tradition. Moscow mathematics has very strong and distinctive features. There are several reasons for this, all of which go back to good and bad aspects of Soviet organization of science. In the twentieth century, there was a veritable galaxy of great mathematicians in Russia, while it so happened that there were only few mathematical centers in which these experts clustered. A major one of these, and perhaps the most influential, was Moscow.

There are three major reasons for the spectacular success of Soviet mathematics:

1. Significant support from the government and the high prestige of science as a profession. Both factors were related to the process of rapid industrialization in the USSR.
2. Doing research in mathematics or physics was one of very few intellectual activities that had no mandatory ideological content. Many would-be computer scientists, historians, philosophers, or economists (and even artists or musicians) became mathematicians or physicists.
3. The Iron Curtain prevented international mobility.

These are specific factors that shaped the structure of Soviet science. Certainly, factors (2) and (3) are more on the negative side and cannot really be called favorable but they essentially came together in combination with the totalitarian system. Nowadays, it would be impossible to find a scientist who would want all of the three factors to be back in their totality. On the other hand, these factors left some positive and long lasting results.

An unprecedented concentration of many bright scientists in few places led eventually to the development of a unique "Soviet school". Of course, mathematical schools in a similar sense were formed in other countries too. An example is the French mathematical school, which has consistently produced first-rate results over a long period of time and where an extensive degree of collaboration takes place. On the other hand, the British mathematical community gave rise to many prominent successes but failed to form a "school" due to a lack of collaborations. Indeed, a school as such is not only a large group of closely collaborating individuals but also a group knit tightly together through student-advisor relationships. In the USA, which is currently the world leader in terms of the level and volume of mathematical research, the level of mobility is very high, and for this reason there are no US mathematical schools in the Soviet or French sense of the term. One can talk not only about the Soviet school of mathematics but also, more specifically, of the Moscow, Leningrad, Kiev, Novosibirsk, Kharkov, and other schools. In all these places, there were constellations of distinguished scientists with large numbers of students, conducting regular seminars. These distinguished scientists were often not merely advisors and leaders, but often they effectively became spiritual leaders in a very general sense.

A characteristic feature of the Moscow mathematical school is that it stresses the necessity for mathematicians to learn mathematics as broadly as they can, rather than focusing on a narrow field in order to get important results as soon as possible.

The Moscow mathematical school is particularly strong in the areas of algebra/algebraic geometry, analysis, geometry and topology, probability, mathematical physics and dynamical systems. The scenarios in which these areas were able to develop in Moscow have passed into history. However, it is possible to maintain and develop the Moscow mathematical tradition in new formats, taking into account modern realities such as globalization and mobility of science. There are three recently created centers—the Independent University of Moscow, the Faculty of Mathematics at the National Research University Higher School of Economics (HSE) and the Center for Advanced Studies at Skolkovo Institute of Science and Technology (SkolTech)—whose mission is to strengthen the Moscow mathematical tradition in new ways. HSE and SkolTech are universities offering officially licensed fulltime educational programs. Mathematical curricula at these universities follow not only the Russian and Moscow tradition but also new global developments in mathematics. Mathematical programs at the HSE are influenced by those of the Independent University of Moscow (IUM). The IUM is not a formal university; it is rather a place where mathematics students of different universities can attend special topics courses as well as courses elaborating the core curriculum. The IUM was the main initiator of the HSE Faculty of Mathematics. Nowadays, there is a close collaboration between the two institutions.

While attempting to further elevate traditionally strong aspects of Moscow mathematics, we do not reproduce the former conditions. Instead of isolation and academic inbreeding, we foster global sharing of ideas and international cooperation. An important part of our mission is to make the Moscow tradition of mathematics at a university level a part of global culture and knowledge.

The "Moscow Lectures" series serves this goal. Our authors are mathematicians of different generations. All follow the Moscow mathematical tradition, and all teach or have taught university courses in Moscow. The authors may have taught mathematics at HSE, SkolTech, IUM, the Science and Education Center of the Steklov Institute, as well as traditional schools like MechMath in MGU or MIPT. Teaching and writing styles may be very different. However, all lecture notes are supposed to convey a live dialog between the instructor and the students. Not only personalities of the lecturers are imprinted in these notes, but also those of students.

We hope that expositions published within the "Moscow lectures" series will provide clear understanding of mathematical subjects, useful intuition, and a feeling of life in the Moscow mathematical school.

Moscow, Russia   Igor M. Krichever
Vladlen A. Timorin
Michael A. Tsfasman
Victor A. Vassiliev

Summary

This book provides a concise but extremely lucid exposition of the basics of algebraic geometry and sheaf theory, seasoned by illuminating examples.

The preprints of this book proved useful to students majoring in mathematics and modern mathematical physics, as well as to professionals in these fields. In particular, this book facilitates grasping notions of supersymmetry theory.

Editor's Preface

It is for more than 50 years now that I wanted to make these lectures—my first love—readily available. The preprints of these lectures [Ma1, Ma2] turned into bibliographic rarities almost immediately. Meanwhile the elements of algebraic geometry became everyday language of working theoretical physicists and the need for concise accessible textbooks only increased. The various (undoubtedly nice) available textbooks are usually too voluminous for anyone who does not aim at becoming a professional algebraic geometer, see, e.g., [GH]; this makes Manin's lecture notes even more appealing.

The methods described in these lectures are currently working tools of theoretical physicists studying subjects that range from high-energy physics (see [Del]), where the Large Hadron Collider still (now is year 2016) struggles to confirm or disprove the supersymmetry of our world (or rather models of it), to solid-state physics, where supersymmetric models already work (see, e.g., the very lucid book [Ef] with a particularly catchy title).

In mathematics, supersymmetry is an integral part of everything related to homotopy and exterior products of differential forms, and hence to (co)homology. In certain subjects, the supersymmetric point of view is unavoidable, e.g., in the study of integrability of certain equations of mathematical physics, see the book [MaG] which, in my opinion, is an extended comment to a remarkable 5-page-long paper by Witten [Wi]. Certain *properties* of solutions to classical equations are also due to supersymmetry, as was first observed also by Witten; for an extremely clear exposition of Witten's approach to this application of supersymmetry, see [GeKr].

For a brief review of new dimensions in geometry ("usual", "arithmetic", and "super" ones), see [MaD].

My definition of superschemes [L0], reported in Moscow at various seminars in 1972, was based on these lectures; they apparently are the briefest and clearest source of the background material needed for studying those aspects of super-symmetry that cannot be reduced to linear algebra. In 1986, Manin wrote a letter allowing me to include a draft of this translation as a chapter in [SoS]; this book is an edited version of this preprint 30/1988-13 of the Department of Mathematics of Stockholm University.

Manin lectured at the same time as Macdonald [M] and Mumford [M1, M2, M3]. Manin's lectures seem to me more lucid and easy, and sprinkled with well-chosen examples illustrating subtleties. Later on, several books elucidating various aspects of the subject were published, e.g., [AM], [Sh0], [E1, E2, E3], [H, Kz, GW]. I particularly recommend [Reid] for the first reading complementary to this book.

The responsibility for footnotes answering the questions of students on which the preprint of these notes was tested, extension of the initial bibliography, and numerous comparisons with the "super" situation (for details, see [SoS]), is mine.

Notation In this book, $\mathbb{N} := \{1, 2, \ldots\}$ and $\mathbb{Z}_+ := \{0, 1, 2, \ldots\}$; $\mathbb{F}_m := \mathbb{Z}/m\mathbb{Z}$. As usual, $\mathbb{Z}_p$ stands for the ring of p-adic integers and $\mathbb{Q}_p$ for the field of rational p-adic numbers. The term "identity" is only applied to relations and maps, the element 1 (sometimes denoted by e) such that $1a = a = a1$ is called a *unit*.

Stockholm, Sweden   D. Leites
May 13, 2016

Acknowledgements I am very thankful to Andrei Iacob, whose editing of my English helped a lot to polish the text, and to NYUAD, where the final editing of this book took place, for stimulating working conditions during 2017.

Author's Preface

During the period 1966–1968, at the Department of Mechanics and Mathematics of Moscow State University, I taught a 2-year-long course in algebraic geometry. The transcripts of the first-year lectures were preprinted [Ma1, Ma2] (and they constitute this book), while those of the second year were published in [Ma3]. These publications retain the style of the lectures, with its pros and cons.

Our goal is to help the reader master the geometric language of commutative algebra. The necessity of presenting separately the algebraic material and later "applying" it to algebraic geometry always discouraged the geometers: O. Zariski and P. Samuel wrote about this very expressively in the preface to their treatise [ZS].

Alexandre Grothendieck's theory of schemes fortunately made possible to remove all boundaries between "geometry" and "algebra": they now manifest as complementary aspects of a whole, like the manifolds and the spaces of functions on them in other geometric theories. From this point of view,

commutative algebra coincides with (more precisely, is functorially dual to) a theory of local geometric objects—the affine schemes.

This book is devoted to deciphering this last claim. I have attempted to consistently explain what sort of geometric representations should be related with concepts such as, say, primary decomposition, modules, and nilpotents. In A. Weil's words, spatial intuition **"is invaluable as long as we are aware of its limitations."** I did strive to take into account both parts of this elegant statement.

Certainly, the accent on geometry did considerably influence the choice of material. In particular, the first chapter should prepare the ground for introducing global objects. For this reason, the section on vector bundles treats, on a "naive" level, constructions that essentially belong already to sheaf theory.

Finally, I did want to introduce notions of category theory as soon as possible; they are not so important in local questions, but play an ever increasing role in the treatment of global questions. I advise the reader to skim through Sect. 1.16 (language of categories) and return to it as needed.

I am absolutely incapable to edit my old texts; if I start doing it, an irresistible desire to throw everything away and completely rewrite the text overcomes me. And doing something new is always more interesting. Therefore I wish to heartily thank D. Leites, who saved me from this frustrating job.

The following list of sources is by no means exhaustive. It may help the reader to quickly grasp the working aspects of the theory: [Sh0, Bb1] (general courses); [M1, M2, Ma3, MaG, S1, S2] (more special questions).[1]

The approach of this book can be extended, to a certain degree, to noncommutative geometry [Kas]. My approach did gradually develop in the direction where the same guiding principle, namely, to construct a matrix with noncommuting elements satisfying only the absolutely necessary commutation relations, proved to be applicable in the ever widening context of **noncommutative** geometries, see [MaT] and also [GK, BM] (extension to operads and further).

Bonn, Germany   Yuri I. Manin
March 22, 2009

---

[1]For further study, read the following: on algebra, [vdW, Lang]; on sheaves, [God, KaS]; on topology, [K, FFG, RF, Bb3, MV]; on number theory, [Sh1, Sh2].

Contents

1 Affine Schemes 1
  1.1   Equations and Rings 1
  1.2   Geometric Language: Points 7
  1.3   Geometric Language, Continuation: Functions on Spectra 11
  1.4   The Zariski Topology on Spec A 16
  1.5   Affine Schemes (A Preliminary Definition) 27
  1.6   Topological Properties of Certain Morphisms
        and the Maximal Spectrum 32
  1.7   Closed Subschemes and the Primary Decomposition 41
  1.8   Hilbert's Nullstellensatz (Theorem on Zeroes) 49
  1.9   Fiber Products 52
  1.10 Vector Bundles and Projective Modules 57
  1.11 The Normal Bundle and Regular Embeddings 66
  1.12 Differentials 70
  1.13 Digression: Serre's Problem and Seshadri's Theorem 75
  1.14 Digression: ζ-function of a Ring 78
  1.15 Affine Group Schemes 84
  1.16 Appendix: The Language of Categories. Representing Functors.... 93
  1.17 Solutions to Selected Problems of Chap. 1 105

2 Sheaves, Schemes, and Projective Spaces 107
  2.1   Basics on Sheaves 107
  2.2   The Structure Sheaf on Spec A 113
  2.3   Ringed Spaces: Schemes 118
  2.4   Projective Spectra 125
  2.5   Algebraic Invariants of Graded Rings 130
  2.6   Presheaves and Sheaves of Modules 139
  2.7   Invertible Sheaves and the Picard Group 146
  2.8   The Čech Cohomology 153
  2.9   Cohomology of the Projective Space 162
  2.10 Serre's Theorem 168

2.11   Sheaves on Proj R and Graded Modules 172
2.12   Applications to the Theory of Hilbert Polynomials 175
2.13   The Grothendieck Group: First Notions 181
2.14   Resolutions and Smoothness 187

References... 195

Index... 201

Chapter 1
Affine Schemes

1.1 Equations and Rings

Studying algebraic equations is an ancient aspect of the mathematical science. In modern times, vogue and convenience dictate us to turn to rings.

1.1.1 Systems of Equations

Let I, J be some sets of indices, let $T = \{T_j\}_{j \in J}$ be indeterminates, let $F = \{F_i \in K[T]\}_{i \in I}$ be a set of polynomials.

A *system X of equations in unknowns T* is a triple (ring K, unknowns T, polynomials F), more exactly and conventionally expressed as

$$F_i(T) = 0, \quad \text{where } i \in I. \tag{1.1}$$

Why the *ground ring* or the *ring of constants* K enters the definition is clear: the coefficients of the polynomials F_i belong to a fixed ring K. We say that the *system X is defined over K*.

What should we take for a *solution* of equations (1.1)?

To say, "a solution of equations (1.1) is a set $t = (t_j)_{j \in J}$ of elements of K such that $F_i(t) = 0$ for all *i*" is too restrictive: we might wish to consider, say, complex roots of equations with real coefficients. The radical resolution of this issue is to consider solutions in **any** ring, and, "for simplicity", in all the rings simultaneously.

To consider solutions of X belonging to a ring L, we should be able to substitute the elements of L into the polynomials F_i with coefficients from K, i.e., we should be able to multiply the elements of L by the elements of K and add the results. For this reason L must be a *K-algebra*.

© The Author(s) 2018
Y.I. Manin, *Introduction to the Theory of Schemes*, Moscow Lectures 1,
https://doi.org/10.1007/978-3-319-74316-5_1

Recall that the set L is said to be a *K-algebra* if L is endowed with structures of a ring and a K-module, connected by the following properties:

1. The multiplication $K \times L \longrightarrow L$ is right and left distributive with respect to addition.
2. $k(l_1 l_2) = (k l_1) l_2$ for any $k \in K$ and $l_1, l_2 \in L$.

A map $f : L_1 \longrightarrow L_2$ is a *K-algebra homomorphism* if f is simultaneously a map of rings and a map of K-modules.

Let 1_L denote the unit of L. To define a K-algebra structure on L, we need a ring homomorphism (embedding of K) $i \colon K \longrightarrow L$ which sends $k \cdot 1_K \in K$ into $i(k) \cdot 1_L$.

1.1.1a Examples

(1) Every ring L is a $\mathbb{Z}$-algebra with respect to the homomorphism $n \longmapsto n \cdot 1_L$ for any $n \in \mathbb{Z}$.
(2) If $K = \mathbb{F}_p$ or $\mathbb{Q}$ and $L = \mathbb{Z}$ or $\mathbb{F}_{p^2}$, then there are no ring homomorphisms $K \longrightarrow L$.

1.1.1b Exercise   Prove claim (2) in Example above.

1.1.2   Solutions of Systems of Equations

A *solution of a system* (1.1) *with values in a K-algebra L* is a set $t = (t_j)_{j \in J}$ of elements of L such that $F_i(t) = 0$ for all $i \in I$. The set of all such solutions is denoted by $X(L)$, and each solution is also called an *L-point of the system of equations X*.

As is clear from Example 1.1.1a (1), for any system of equations with *integer* coefficients, we may consider its solution in *any* commutative ring.

Two systems of equations X and Y for the same unknowns given over a ring K are said to be *equivalent*, or, to be more precise, *equivalent over K* (and we write $X \sim Y$) if $X(L) = Y(L)$ for any K-algebra L.

Among the systems of equations equivalent to a given system X, there exists the "biggest" one: for the left-hand sides of this "biggest" system one takes the left-hand sides of Eq. (1.1) that generate the ideal (F) in $K[T]_{j \in J}$.

The *coordinate ring* of the variety $X = V(F)$ is

$$K(X) = K[T]/(F).$$

1.1.2a Proposition

(1) *The system of equations whose left-hand sides are elements of the ideal (F) is the largest system of equations equivalent to (F).*
(2) *There is a one-to-one correspondence $X(L) \longleftrightarrow \mathrm{Hom}_K(K(X), L)$, where Hom_K denotes the set of K-algebra homomorphisms.*

Proof

(1) Let P be the ideal in $K[T]$, where $T = (T)_{j \in J}$, generated by the left-hand sides of the system of Eq. (1.1). It is easy to see that the system of equations obtained by equating all elements of P to zero is equivalent to our system, call it (X). At the same time, this larger system is maximal in the sense that if we add to it any equation $f = 0$ not initially contained in it, then we get a system of equations that is not equivalent to (X). Indeed, take $L = K[T]/P$. In the ring L, the element t, where $t_j \cong T'_j \pmod P$, is a solution of the initial system (X), whereas $f(t) \neq 0$ because $f \notin P$.

(2) Let $t = (t_j)_{j \in J} \in X(L)$. There exists a K-algebra homomorphism $K[T] \longrightarrow L$ which coincides on K with the structure homomorphism $K \longrightarrow L$ and sends T_j to t_j. By the definition of $X(L)$, the ideal P lies in the kernel of this homomorphism, so we can consider a through homomorphism $A = K[T]/P \longrightarrow L$.

Conversely, any K-algebra homomorphism $A \longrightarrow L$ uniquely determines a through homomorphism $K[T] \longrightarrow A \longrightarrow L$. Let t_j be the image of T_j under this through homomorphism; then $t = (t_j)_{j \in J} \in X(L)$, because all elements of P vanish under this homomorphism.

It is easy to check that the maps $X(L) \longleftrightarrow \mathrm{Hom}_K(K(X), L)$ we constructed are inverses of each other. $\square$

For a non-zero K-algebra L, a system of equations X over a ring K is said to be *consistent over L* if $X(L) \neq \emptyset$, and *inconsistent* otherwise.[1] Proposition 1.1.2a shows that the system of equations X is inconsistent only if its coordinate ring, or rather, algebra, $K(X)$, reduces to zero, in other words, only if $1 \in (X)$.

1.1.3 Examples from Arithmetics

1.1.3a. The Language of Congruences

Let n be an integer of the form $4m + 3$. The classical proof of the fact that n is not representable as a sum of two perfect squares is as follows: if it were, the congruence

$$T_1^2 + T_2^2 \equiv 3 \pmod 4$$

would have been solvable, whereas a simple case-by-case checking (set $T_1 = 4a + r_1$ and $T_2 = 4b + r_2$ and consider the eight distinct values of (r_1, r_2)) establishes that this is not the case.

---

[1] In high school, college, and even in university, one often omits "over L" in the definition of consistency, thus declaring systems that are only partly inconsistent (over some classes of rings) as totally inconsistent.

From our point of view this argument reads as follows: consider the system consisting of a single equation

$$T_1^2 + T_2^2 = n, \text{ where } K = \mathbb{Z}.$$

The reduction modulo 4, i.e., the map $\mathbb{Z} \longrightarrow \mathbb{Z}/4\mathbb{Z}$, determines, in its turn, a map $X(\mathbb{Z}) \longrightarrow X(\mathbb{Z}/4\mathbb{Z})$ and, if $X(\mathbb{Z}) \neq \emptyset$, then $X(\mathbb{Z}/4\mathbb{Z}) \neq \emptyset$, which is not the case.

More generally, in order to study $X(\mathbb{Z})$ for any system of equations X with integer coefficients, we can consider the sets $X(\mathbb{Z}/m\mathbb{Z})$ for any m and try to deduce from this consideration some information on $X(\mathbb{Z})$. Usually,

if $X(L) = \emptyset$ for a nontrivial (i.e., $1 \neq 0$) field L, then $X(\mathbb{Z}) = \emptyset$.

In practice, one usually tests $L = \mathbb{R}$ and the finite fields $L = \mathbb{F}_p(:= \mathbb{Z}/(p))$ for all prime p.

A series of the deepest results of the theory of Diophantine equations are related with the problem: when is the converse statement true? A prototype of these results is

1.1.3a.i Theorem (Legendre's Theorem, [BSh]) *Consider the equation*

$$a_1 T_1^2 + a_2 T_2^2 + a_3 T_3^2 = 0, \quad K = \mathbb{Z}. \tag{1.2}$$

If $X(\mathbb{Z}) = \{(0,0,0)\}$, then $X(L) = \{(0,0,0)\}$ for at least one of the rings $L = \mathbb{R}$ or $L = \mathbb{Z}/m\mathbb{Z}$, where $m > 1$.

1.1.3b. Equations in Prime Characteristic

Consider the equation

$$0 \cdot T + 2 = 0, \quad K = \mathbb{Z}.$$

Clearly, $X(L) = \emptyset$ if $2 \cdot 1_L \neq 0$ and $X(L) = L$ if $2 \cdot 1_L = 0$.

This example is manifestly rather artificial; still, we often encounter the like of it in "arithmetic geometry".

1.1.3c. On Usefulness of Complexification

When studying $X(\mathbb{R}) \subset \mathbb{R}^n$ (for the case where $K = \mathbb{R}$, and the number of unknowns is equal to n), it is expedient to pass to the complexification of $X(\mathbb{R})$, i.e., to $X(\mathbb{C})$. Since $\mathbb{C}$ is algebraically closed, it is usually easier to study $X(\mathbb{C})$ than $X(\mathbb{R})$; this often constitutes the first stage of the investigation even when we are primarily interested in purely real problems. The following example is illuminating.

1.1.3c.i Theorem (Harnak's Theorem) *Let $F(T_0, T_1, T_2)$ be a form (i.e., a homogeneous polynomial) of degree d with real coefficients. Let $X(\mathbb{R})$ be the curve in $\mathbb{R}P^2$, the real projective plane, singled out by the equation $F = 0$. Then the number of connected components of $X(\mathbb{R})$ does not exceed $\frac{1}{2}(d-1)(d-2) + 1$.*

The proof uses the embedding $X(\mathbb{R}) \longrightarrow X(\mathbb{C})$. For simplicity, let us confine ourselves to the case of a nonsingular $X(\mathbb{C})$, i.e., to the case where $X(\mathbb{C})$ is a compact orientable 2-dimensional manifold. Its genus—the number of handles— is then equal to $\frac{1}{2}(d-1)(d-2)$. In Fig. 1.1, $d = 3$ and $X(\mathbb{C})$ is a torus. The proof is based on the following two statements:

First, complex conjugation acts continuously on $X(\mathbb{C})$, and $X(\mathbb{R})$ is the set of its fixed points.

Second, if one cuts $X(\mathbb{C})$ along $X(\mathbb{R})$, then $X(\mathbb{C})$ splits into precisely two pieces, cf. Fig. 1.2 (like, for $d = 1$, the Riemann sphere—the compactified real plane— splits when cut along the real axis). Now, routine topological considerations give Harnak's estimate, see, e.g., [Ch].

1.1.3d. The Algebra of Mathematical Logic in Geometric Terms

(For mathematical logic from an algebraist's point of view, see [Ma4].) A *Boolean ring* is any ring R (with unit 1) such that $P^2 = P$ for any $P \in R$. Clearly,

$$P + Q = (P + Q)^2 = P^2 + PQ + QP + Q^2 = P + PQ + QP + Q \qquad (1.3)$$

implies $PQ + QP = 0$. Since R is commutative by definition, $2PQ = 0$; moreover,

$$2P = P + P = P^2 + P^2 = 0$$

implies $P = -P$. Therefore, every Boolean ring is a commutative ring over $\mathbb{F}_2$.

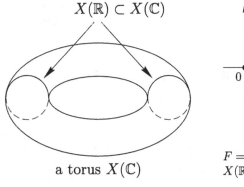

$X(\mathbb{R}) \subset X(\mathbb{C})$

a torus $X(\mathbb{C})$

Fig. 1.1

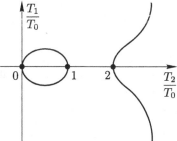

$F = T_0 T_1^2 - T_2(T_2 - T_0)(T_2 - 2T_0)$
$X(\mathbb{R})$ on the projective plane

Fig. 1.2

It is not difficult to show that every prime ideal of a Boolean algebra R is maximal, and therefore every element $P \in R$ can be viewed as an $\mathbb{F}_2$-valued function on $\operatorname{Spec} R$.

Given two statements, P and Q, each either true of false, define their *sum* and *product* by setting

$$P + Q := (P \vee Q) \wedge (\overline{P} \vee \overline{Q}), \quad PQ := P \wedge Q,$$

where the bar stands for negation, $\wedge$ for conjunction and $\vee$ for disjunction. With respect to the above operations the empty statement $\emptyset$ is the zero, and $\overline{\emptyset}$ is the unit. Clearly, $P^2 = P$ and $2P = P + P = 0$ for any P.[2]

1.1.4   Summary

We have established the *equivalence of two languages*: that of systems of equations (which is used in concrete calculations) and that of rings and their morphisms. More exactly, we have established the following equivalences:

$$\left\{ \begin{array}{c} \text{A system of equations } X \\ \text{over a ring } K \\ \text{for unknowns } \{T_j \mid j \in J\}. \end{array} \right\} \Longleftrightarrow \left\{ \begin{array}{c} \text{A } K\text{-algebra } K(X) \text{ with a} \\ \text{system of generators} \\ \{t_j \mid j \in J\}. \end{array} \right\}$$

$$\left\{ \begin{array}{c} \text{A solution of the system} \\ \text{of equations } X \\ \text{in a } K\text{-algebra } L. \end{array} \right\} \Longleftrightarrow \left\{ \begin{array}{c} \text{A } K\text{-algebra homomorphism} \\ K(X) \longrightarrow L. \end{array} \right\}$$

Finally, notice that when using the language of rings there is no need to consider a fixed system of generators $t = (t_j)_{j \in J}$ of $K(X) = K[T]/(X)$; rather, we should identify the systems of equations obtained from one another by an invertible change of unknowns. Every generator of $K(X)$ plays the role of one of the "unknowns", and the value this unknown takes at a given solution of the system of equations coincides with its image in L under the corresponding homomorphism.

1.1.5 Exercises
(1) The equation $2T - 4 = 0$ is equivalent to the equation $T - 2 = 0$, if and only if 2 is invertible in K.

---

[2]**Question.** The wedge sign, although seemingly accidental, suggests a deeper analogy of mathematical logic with superschemes. I know of only one attempt to investigate this analogy: [Ru]. Is it possible to use odd indeterminates to encode conjunction and disjunction, and interpret negation in terms of them?

(2) The equation $(T - 1)^2 = 0$ is not equivalent to the equation $T - 1 = 0$.

(3) Let the system of equations $\{F_i(X) = 0 \mid i \in I\}$, where $X = (X_j)_{j \in J}$, be inconsistent. Then it has a finite inconsistent subsystem.

(4) Let $T_1, \ldots, T_n$ be indeterminates, $s_i(T)$ the i-th elementary symmetric polynomial in them, and $p_i(T) := \sum_{1 \leq j \leq n} T_j^i$. Determine the rings of constants over which the following systems of equations are equivalent:

$$X_1: \quad s_i(T) = 0 \quad \text{for} \quad i = 1, \ldots, k \leq n,$$

$$X_2: \quad p_i(T) = 0 \quad \text{for} \quad i = 1, \ldots, k \leq n.$$

Hint Use *Newton's formulas.*[3]

(5) Any system of equations over a ring K in a finite number of unknowns $T_1, \ldots, T_n$ is equivalent to a finite system of equations if and only if the ring $K[T_1, \ldots, T_n]$ is *Noetherian.*[4]

(6) Let X be a system of equations over K and A the ring corresponding to X. The maps $L \longmapsto X(L)$ and $L \longmapsto \mathrm{Hom}_K(A, L)$ determine covariant functors on the category Algs_K of K-algebras with values in the category Sets. Verify that Proposition 1.1.2a determines an isomorphism of these functors.

1.2 Geometric Language: Points

Let, as above, K be the ground ring, and X a system of equations over K in the unknowns $T_1, \ldots, T_n$.

For any K-algebra L, we realize the set $X(L)$ as a "graph" in L^n, the coordinate space over L. The points of this graph are *solutions of the system of equations* (1.1). Taking into account the results of the preceding section, we introduce the following definition.

---

[3]Recall that *Newton's identities*, aka *Newton–Girard formulas*, valid for all $n \geq 1$ and $k \geq 1$, are

$$ks_k(T) = \sum_{1 \leq i \leq k} (-1)^{i-1} s_{k-i}(T) p_i(T).$$

[4]A ring is said to be *Noetherian* if it satisfies any of the following equivalent conditions [Lang]:

(1) Any set of its generators contains a finite subset.

(2) Any ascending chain of its ideals (say, only left ideals if rings are non-commutative) stabilizes.

1.2.1 The Points of a K-algebra A

The *points* of a K-algebra A with values in a K-algebra B (or just B-*points* of A) are the K-homomorphisms $A \longrightarrow B$. A B-point of A is called *geometric* if B is a field.

1.2.1a Example (This example shows where the motivation for using the term "point" for a homomorphism comes from.) Let K be a field, V an n-dimensional vector space over K. Let us show that the set of point of V is in a one-to-one correspondence with the set of maximal ideals of the ring $K[x]$, of polynomial functions on V, where $x = (x_1, \ldots, x_n)$.

Every point (vector) of V is a linear functional on V^*, the space of linear functionals on V with values in K and, as one can easily show [Lang], this linear functional can be extended to an algebra homomorphism

$$S^{\cdot}(V^*) \xrightarrow{\simeq} K[x_1, \ldots, x_n] \longrightarrow K,$$

where $V^* = \mathrm{Span}(x_1, \ldots, x_n)$ and $S^{\cdot}(V^*)$ is the symmetric algebra of the space V^*. This homomorphism can be canonically identified with a point (vector) of V.

The next step is to distinguish the properties of a fixed algebra A from those of the variable algebra B, namely, instead of homomorphisms $h: A \longrightarrow B$ we consider their kernels, which are ideals in A.

1.2.2 Spectrum

The kernel of a homomorphism $A \longrightarrow B$ corresponding to a geometric point is, clearly, a prime ideal (to be defined shortly). There are many reasons why one should confine oneself to prime ideals instead of the seemingly more natural maximal ones; in the following sections we give these reasons.

A proper ideal $\mathfrak{p}$ of a commutative ring A is said to be *prime* if $A/\mathfrak{p}$ is an integral domain, i.e., has no zero divisors (and if we do not forbid $1 = 0$, the zero ring cannot be an integral domain). Equivalently, $\mathfrak{p}$ is *prime* if $\mathfrak{p} \neq A$ and

$$a \in A, \quad b \in A, \quad ab \in \mathfrak{p} \Longrightarrow \text{ either } a \in \mathfrak{p} \text{ or } b \in \mathfrak{p}.$$

The set of all prime ideals of A is said to be the (prime) *spectrum of A* and is denoted by $\mathrm{Spec}\, A$. The elements of $\mathrm{Spec}\, A$ are called its *points*.

In what follows, we enrich the set $\mathrm{Spec}\, A$ with additional structures making it into a topological space rigged with a sheaf of rings: this will lead to the definition of an affine scheme. Schemes, i.e., topological spaces with sheaves, locally isomorphic to affine schemes, are the main actors in algebraic geometry.

Embarking upon the study of spectra we have to verify, first of all, that there is indeed what to study, i.e., that they are nontrivial.

1.2.3 Theorem *If $A \neq 0$, then* Spec $A \neq \emptyset$.

In the proof of this theorem we need the following:

1.2.4 Lemma (Zorn's Lemma) *Any partially ordered set containing upper bounds for every chain (that is, every totally ordered subset) necessarily contains at least one maximal element.*

For a proofs of Zorn's lemma, see, e.g., [K] or [Hs], where it is proved together with its equivalence to the Axiom of Choice, the well-ordering principle, and several other statements.[5]   □

Any ordered set satisfying the condition of Zorn's lemma is called *inductive*.

Proof of Theorem 1.2.3 Denote by M the set of all ideals of A different from A. Since M contains (0), it follows that $M \neq \emptyset$. The set M is partially ordered with respect to inclusion. Take an arbitrary linearly ordered subset $\{\mathfrak{p}_\alpha\}_{\alpha \in \Lambda}$ in M. Then $\bigcup_\alpha \mathfrak{p}_\alpha$ is also an ideal of A (mind the linear order) and this ideal is different from A (because the unit element does not belong to $\bigcup_\alpha \mathfrak{p}_\alpha$). Therefore, M is an inductive set. Denote by $\mathfrak{p}$ its maximal element; it is a maximal ideal, and therefore a prime one: in $A/\mathfrak{p}$, every non-zero ideal, in particular, all principal ideals, coincide with $A/\mathfrak{p}$. Therefore, every non-zero element of $A/\mathfrak{p}$ is invertible and $A/\mathfrak{p}$ is a field.   □

1.2.4a Corollary *Every prime ideal is contained in a maximal ideal.*

Theorem 1.2.3 implies, in particular, that the spectrum of every non-zero ring A possesses geometric points (e.g., any homomorphism $A \longrightarrow A/\mathfrak{p}$, where $\mathfrak{p} \subset A$ is a maximal ideal, is one of them).

1.2.5   The Center of a Geometric Point

The *center* of a geometric point $A \longrightarrow L$ is its kernel, regarded as an element of Spec A. Let $k(x)$ be the *field of fractions*[6] of the ring $A/\mathfrak{p}_x$, where $\mathfrak{p}_x \subset A$ is the ideal corresponding to $x \in $ Spec A.

1.2.5a Proposition *The geometric L-points of a K-algebra A with center in $x \in$ Spec A are in a one-to-one correspondence with the K-homomorphisms $k(x) \longrightarrow L$.*

Proof Indeed, L is a field, so any homomorphism $A \longrightarrow L$ factorizes as

$$A \longrightarrow A/\mathfrak{p}_x \longrightarrow k(x) \longrightarrow L.$$

---

[5]For interesting new additions to the list of equivalent statements, see [Bla].

[6]The *field of fractions* or *field of quotients* of a given ring is the smallest field in which the ring can be embedded. It will be constructed explicitly in what follows.

The first two arrows of this sequence are determined once and for all.   □

1.2.6 Examples

(a) Let K be a perfect[7] field, L an algebraically closed field containing K; let A be a K-algebra, $x \in \text{Spec}\, A$, and $\mathfrak{p}_x$ the corresponding ideal.

If $\deg x := [k(x) : K] < \infty$, then, by Galois theory, there are exactly $\deg x$ geometric L-points with center in x.

If the field $k(x)$ is not algebraic over K, and L contains sufficiently many *transcendental* elements over K, then there may be infinitely many geometric L-points with center in x.

Here is a rather special case.

(b) The set of geometric $\mathbb{C}$-points of $\mathbb{R}[T]$ is the complex line $\mathbb{C}$. The set $\text{Spec}\,\mathbb{R}[T]$ is the union of the zero ideal (0) and the set of all monic polynomials irreducible over $\mathbb{R}$. Every such polynomial of degree 2 has two complex conjugate roots corresponding to two distinct geometric points, see Fig. 1.3.

In general, for any perfect field K, the geometric points of the K-algebra $K[T]$ with values in the algebraic closure $\overline{K}$ are simply the elements of $\overline{K}$, and their centers are irreducible polynomials over K, i.e., the sets consisting of all elements of $\overline{K}$ conjugate over K to one of them.   □

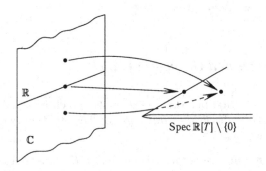

Spec $\mathbb{R}[T] \setminus \{0\}$

Fig. 1.3

---

[7]Recall that a field K of characteristic p is said to be *perfect* if $K^p := \{x^p \mid x \in K\}$ coincides with K. (The symbol K^p is also used to designate $\underbrace{K \times \cdots \times K}_{p\ \text{times}}$.)

1.2.7 A Duality: Unknowns ⟷ Coefficients

When considering $\operatorname{Spec} A$ we may forget, if needed, that A is a K-algebra; every ideal of A is stable under multiplication by the elements of K. When, however, we are interested in geometric points (or, more generally, in arbitrary L-points), the reference to K is essential, because we have to consider K-homomorphisms $A \longrightarrow L$. Arbitrary ring homomorphisms are, clearly, $\mathbb{Z}$-homomorphisms; therefore this "absolute" case may be considered as a specialization of a "relative" case, the one over K.

For the systems of equations, the passage to the absolute case means that we ignore the distinction between "unknowns" and "coefficients" and are allowed to vary the values of both. More precisely, consider a system of equations in which the jth equation is

$$\sum_k a_k^{(j)} x^k = 0, \quad \text{where } x^k \text{ runs over the monomials in our indeterminates.}$$

Generally, $a_k^{(j)}$ are fixed elements of a ring of constants K, whereas "passage to the absolute case" means that we now *write a generating set of all relations between the $a_k^{(j)}$ over $\mathbb{Z}$, and add it to our initial system of equations, binding the x_i and the $a_k^{(j)}$ together. And after that we may specialize the coefficients as well, retaining only the relations between them.*

1.2.8 Exercise (A Weak Form of Hilbert's Nullstellensatz (Theorem on Zeros), cf. Theorem 1.8.4) Over the ring K, consider a system of equations $\{F_i(T) = 0\}$, where $T = (T_j)_{j \in J}$. Then either this system has a solution with values in a field, or there exist polynomials $G_i \in K[T]$ (finitely many of which are $\neq 0$) such that

$$\sum_i G_i F_i = 1.$$

Hint Apply Theorem 1.2.3 to the ring corresponding to the system of equations.

1.3 Geometric Language, Continuation: Functions on Spectra

1.3.1 Functions on Spectra

Let X be a system of equations over K in unknowns $T = (T_j)_{j \in J}$. Every solution of X in a K-algebra L, i.e., an element of $X(L)$, evaluates the T_j; let $\{t_j \in L\}_{j \in J}$ be the corresponding values. Therefore, it is natural to consider T_j

as a function on $X(L)$ with values in L. Clearly, this function depends only on the class of T_j modulo the ideal generated by the left-hand sides of X. This class is an element of the K-algebra $K(X) := K[T]/(X)$ associated with X, and generally all elements of $K(X)$ are functions on $X(L) = \mathrm{Hom}_K(K(X), L)$: for every $\varphi : K(X) \longrightarrow L$ and $f \in K(X)$, "the value of f at φ" is by definition $\varphi(f)$.

The classical notation of functions is not well suited to reflect the fundamental duality that increasingly manifests itself in modern mathematics:

"The space $\longleftrightarrow$ the ring of functions on the space"

or, symmetrically, with a different emphasis:

"The ring $\longleftrightarrow$ the spectrum of its ideals of certain type".

When applied to $\mathrm{Spec}\, A$, this duality leads one to regard any element $f \in A$ as a function on $\mathrm{Spec}\, A$. Let $x \in \mathrm{Spec}\, A$ and let $\mathfrak{p}_x$ be the corresponding ideal. Then, by definition, $f(x) = f \pmod{\mathfrak{p}_x}$ and we consider $f(x)$ as belonging to the field of fractions $k(x)$ of the ring $A/\mathfrak{p}_x$.

1.3.1a Convention In what follows, when speaking about *functions* on $\mathrm{Spec}\, A$ we will always mean the elements of A.

Thus, to every point $x \in \mathrm{Spec}\, A$ there is assigned **its own** field $k(x)$, and the values of the functions on $\mathrm{Spec}\, A$ belong to these fields.

In Fig. 1.4, I tried to plot the first five integers (0, 1, 2, 3, 4) considered as functions on $\mathrm{Spec}\, \mathbb{Z}$. The picture is not too convincing; besides, for various reasons that lie beyond the scope of these lectures, the "line" over the field $\mathbb{Z}/p\mathbb{Z}$—the "vertical axis" over the point (p)—should be drawn "coiled into a ring", i.e., the points of this "line" should form the vertices of a regular p-gon, which does not make the task of the artist simple.

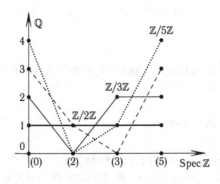

Fig. 1.4

To distinct elements of A there may correspond same functions on the spectrum; their difference represents the zero function, i.e., their difference belongs to $\bigcap_{x \in \operatorname{Spec} A} \mathfrak{p}_x$. Clearly, all nilpotents are contained in this intersection. Let us prove the opposite inclusion. For this, we need a new notion.

The set $\mathfrak{n}(A)$ of all nilpotent elements of a ring A is an ideal (as is easy to see), this ideal is called the *nilradical* of A.

1.3.2 Theorem *Any function that vanishes at all the points of* $\operatorname{Spec} A$ *is represented by a nilpotent element of* A. *In other words,*

$$\mathfrak{n}(A) = \bigcap_{\mathfrak{p} \in \operatorname{Spec} A} \mathfrak{p}.$$

Proof It suffices to establish that, for every non-nilpotent element, there exists a prime ideal which does not contain this element.

Let $h \in A$ and $h^n \neq 0$ for any positive integer n. Let M be the set of all ideals of A that do not contain h^m for any $m \in M$; then $M \neq \emptyset$, because M contains (0). The inductive property of M can be proved as in Theorem 1.2.3. Let $\mathfrak{p}$ be a maximal element of M. We prove that $\mathfrak{p}$ is prime.

Let $f, g \in A$ and $f, g \notin \mathfrak{p}$. We claim that $fg \notin \mathfrak{p}$. Indeed,[8] $\mathfrak{p} + (f) \supset \mathfrak{p}$ and $\mathfrak{p} + (g) \supset \mathfrak{p}$ (strict inclusions). Since $\mathfrak{p}$ is maximal in M, we see that $h^n \in \mathfrak{p} + (f)$ and $h^m \in \mathfrak{p} + (g)$ for some $m, n \in \mathbb{N}$. Hence, $h^{n+m} \in \mathfrak{p} + (fg)$; but $h^{n+m} \notin \mathfrak{p}$, and hence $fg \notin \mathfrak{p}$. Thus, $\mathfrak{p}$ is prime. $\square$

This result might create the wrong impression that in a geometric picture there is no room for nilpotents.[9] On the contrary, nilpotents provide us with an adequate language in which to describe differential-geometric notions like "tangency", "multiplicity of an intersection", "infinitesimal deformation", "fibers of a map at points where regularity is violated".

1.3.3 Examples
(1) *A multiple intersection point.* In the affine plane over $\mathbb{R}$, consider the parabola $T_1 - T_2^2 = 0$ and the straight line $T_1 - t = 0$, where $t \in \mathbb{R}$ is a parameter.

Their intersection is given by the system of equations

$$\begin{cases} T_1 - T_2^2 &= 0, \\ T_1 - t &= 0, \end{cases}$$

---

[8] Here (f) denotes the ideal generated by an element f. The notation $(f_1, \ldots, f_n)$ stands—in this context—for the finitely generated ideal, not the vector $(f_1, \ldots, f_n) \in A^n$.

[9] Instances of this kind are illustrated by Examples 1.3.3 and in the next section, to say nothing of the main body of any (reasonable) textbook on supermanifold theory (e.g., [Del, SoS]), where odd indeterminates, especially odd parameters of representations of Lie supergroups and Lie superalgebras, are the heart (and the *soul*, as B. de Witt might had put it) of the matter.

to which there corresponds the ring $A_t = \mathbb{R}[T_1, T_2]/(T_1 - T_2^2, \ T_1 - t)$. An easy calculation shows that

$$A_t \cong \begin{cases} \mathbb{R} \times \mathbb{R}, & \text{if } t > 0, \\ \mathbb{R}[T]/(T^2), & \text{if } t = 0, \\ \mathbb{C}, & \text{if } t < 0. \end{cases}$$

The *geometric* $\mathbb{R}$-*points* of A_t are as follows: two for $t > 0$, one for $t = 0$, and none for $t < 0$.

The *geometric* $\mathbb{C}$-*points*: there are always two of them, except for $t = 0$ (the case of tangency).

In order to be able to state that over $\mathbb{C}$ there are always two intersection points if proper *multiplicities* are ascribed to them, we have to assume that at $t = 0$ the multiplicity of the intersection point is equal to 2.

Observe that $\dim_{\mathbb{R}} A_t = 2$ regardless of the value of t. The equality

$$\dim_{\mathbb{R}} A_t = \text{the number of intersection points (counting multiplicities)}$$

is not accidental, and we will prove the corresponding theorem when we introduce the projective space, which will enable us also to take into account the points that escaped to infinity.

A singularity, such as the coincidence of the intersection points corresponding to tangency, creates nilpotents in A_0.

(2) *One-point spectra.* Let $\operatorname{Spec} A$ consist of a single point corresponding to an ideal $\mathfrak{p} \subset A$. Then $A/\mathfrak{p}$ is a field and $\mathfrak{p}$ consists of nilpotents. The ring A is Artinian,[10] hence Noetherian, and the standard arguments show that $\mathfrak{p}$ is a nilpotent ideal.

Indeed, let $f_1, \ldots, f_n$ be generators of $\mathfrak{p}$ and $f_i^m = 0$ for $1 \leq i \leq n$. Then

$$\prod_{j=1}^{mn} \left(\sum_{i=1}^{n} a_{ij} f_i \right) = 0$$

for any $a_{ij} \in A$, where $1 \leq i \leq n$ and $1 \leq j \leq mn$, because every monomial of the product contains at least one of the f_i raised to some power $\geq m$.

Recall that the *length of a module* M is the length r of the longest filtration

$$M = M_1 \supsetneq M_2 \supsetneq \cdots \supsetneq M_r = 0. \tag{1.4}$$

---

[10]Recall that the ring is said to be *Artinian* if it satisfies *the descending chain condition* (DCC) *on ideals.*

The filtration (1.4) of a given module M is said to be *simple* if each module M_i/M_{i+1} is simple, and the module is said to be *simple* if it contains no proper submodules (different from $\{0\}$ and itself). A module is said to be of *finite length* if either it is $\{0\}$, or admits a simple finite filtration.

Returning to our case we thus see that $\mathfrak{p}^{mn} = 0$. In the filtration

$$A \supset \mathfrak{p} \supset \mathfrak{p}^2 \supset \cdots \supset \mathfrak{p}^{mn} = (0),$$

the quotients $\mathfrak{p}^i/\mathfrak{p}^{i+1}$ are finite-dimensional vector spaces over the field $A/\mathfrak{p}$. Therefore, as a module over itself, A is of finite length. In intersection theory, the length of a local ring A plays the role of the multiplicity of the only point of $\operatorname{Spec} A$, as we have just seen. The multiplicity of the only point of $\operatorname{Spec} A$ is equal to 1 if and only if A is a field.

(3) *Differential neighborhoods. Jets.* Let $x \in \operatorname{Spec} A$ be a point, $\mathfrak{p}_x$ the corresponding ideal. In differential geometry, and even in freshmen calculus courses, we often have to consider the m-th differential neighborhood of x, i.e., take into account not only the values of functions, but also the values of their derivatives of order $\leq m$, in other words, consider the m-jets of functions at x. This is equivalent to considering the function's Taylor series expansion in which the infinitesimals of order greater than m are disregarded, i.e., set equal to 0.

Algebraically, this means that we consider the class $f \pmod{\mathfrak{p}_x^{m+1}}$. The elements of $\mathfrak{p}_x$ are infinitesimals of order ≥ 1, and, in the ring $A/\mathfrak{p}_x^{m+1}$, they turn into nilpotents.

In what follows we will see that the interpretation of $\operatorname{Spec} A/\mathfrak{p}_x^{m+1}$ as the differential neighborhood of x is only natural when $\mathfrak{p}_x$ is maximal. In the general case of a prime but not maximal $\mathfrak{p}_x$, this intuitive interpretation cannot *guide* us, but is still useful.

(4) *Reduction modulo p^N.* When considering Diophantine equations or, equivalently, the quotients of the ring $\mathbb{Z}[T_1, \ldots, T_n]$, one often makes use of the reduction modulo powers of a prime, see Sect. 1.1.3. This immediately leads to nilpotents, and we see that from the algebraic point of view this process does not differ from the consideration of differential neighborhoods in the above example.

(The congruence $3^5 \cong 7 \pmod{5^3}$ means that "at the point (5) the functions 3^5 and 7 and their derivatives of order ≤ 2 coincide". This language does not look too extravagant in the number theory after the introduction of p-adic numbers by Hensel.)

1.3.4 Exercises

(1) Let $\mathfrak{a}_1, \ldots, \mathfrak{a}_n \subset A$ be ideals. Prove that

$$V(\mathfrak{a}_1 \cdots \mathfrak{a}_n) = V(\mathfrak{a}_1 \cap \cdots \cap \mathfrak{a}_n).$$

(2) Let $f_1, \ldots, f_n \in A$, and let $m_1, \ldots, m_n$ be positive integers. If $(f_1, \ldots, f_n) = A$, then $(f_1^{m_1}, \ldots, f_n^{m_n}) = A$.

(3) The elements $f \in A$ that do not vanish at any point of Spec A are invertible.

1.4   The Zariski Topology on Spec A

The minimal natural condition for a topology to be compatible with "functions" is that the set of zeroes of any function (which are only polynomial ones in these lectures) should be closed. The topology on Spec A that satisfies this criterion is called the *Zariski topology*. To describe it, for any subset $E \subset A$, denote the variety singled out by E by

$$V(E) = \{x \in \operatorname{Spec} A \mid f(x) = 0 \text{ for any } f \in E\}.$$

1.4.1 Lemma *The sets $V(E)$, where $E \subset A$, constitute the family of closed sets in a topology of Spec A called the Zariski topology.*

Proof Since $\emptyset = V(1)$ and Spec $A = V(\emptyset)$, it suffices to verify that $V(E)$ is stable under taking finite unions and arbitrary intersections. This follows from the next statement.

1.4.1a Exercise Denote $E_1 E_2 := \{fg \mid f \in E_1, g \in E_2\}$.

Then $V(E_1) \cup V(E_2) = V(E_1 E_2)$ and $\bigcap_{i \in I} V(E_i) = V\left(\bigcap_{i \in I} E_i\right)$ for any I.   □

Using Theorem 1.3.2, we can describe the set of functions that vanish on $V(E)$. Obviously, it contains all elements of the ideal (E) generated by E, as well as all elements $f \in A$ such that $f^n \in (E)$ for some n. It turns out that this is all.

The *radical* $\mathfrak{r}(I)$ *of the ideal* $I \subset A$ is the set (actually, an ideal in A)

$$\mathfrak{r}(I) = \{f \in A \mid \text{ there exists } n \in \mathbb{N} \text{ such that } f^n \in I\}. \tag{1.5}$$

In particular, the nilradical $\mathfrak{n}(A)$ is the radical of the zero ideal. An ideal that coincides with its own radical is called a *radical ideal*.

1.4.2 Theorem *If $f(x) = 0$ for any $x \in V(E)$, then $f \in \mathfrak{r}((E))$.*

Proof The condition

$$f(x) = 0 \text{ for all } x \in V(E)$$

means that $f \in \bigcap_{\mathfrak{p}_x \subset \operatorname{Spec} A/(E)} \mathfrak{p}_x$, i.e., every element $f \pmod{(E)} \in A/(E)$ belongs to the intersection of all prime ideals of $A/(E)$. Therefore, thanks to Theorem 1.3.1, $f^n \pmod{(E)} = 0$ for some n. This proves the statement of Theorem 1.4.2.   □

1.4.2a Corollary *The map* $I \longmapsto V(I)$ *establishes a one-to-one correspondence between the radical ideals of the ring A and the closed subsets of its spectrum.*

Proof immediately follows from Theorem 1.3.1.   □

The topology of the spaces Spec A is very non-classical, in the sense that it is very non-Hausdorff (non-separable). We consider certain typical phenomena specific for algebraic geometry.

1.4.3   Non-closed Points

Let us find the closure of a given point $x \in \operatorname{Spec} A$. We have

$$\overline{\{x\}} = \bigcap_{E \subset \mathfrak{p}_x} V(E) = V\left(\bigcup_{E \subset \mathfrak{p}_x} E \right) = V(\mathfrak{p}_x) = \{\, y \in \operatorname{Spec} A \mid \mathfrak{p}_y \supset \mathfrak{p}_x \}.$$

In other words,

$$\overline{\{x\}} \cong \operatorname{Spec}(A/\mathfrak{p}_x), \text{ and}$$

only the points corresponding to the maximal ideals are closed.

The specific relation $y \in \overline{\{x\}}$ between points is sometimes expressed by saying that y is a *specialization* of x; this is equivalent to the inclusion $\mathfrak{p}_x \subset \mathfrak{p}_y$. If A has no zero divisors, then $\{0\} \in \operatorname{Spec} A$ is the point *whose closure coincides with the whole spectrum.*

Therefore, Spec A is stratified: the closed points are on the highest level, the preceding level is occupied by the points whose specializations are closed, and so on, the i-th level (from above) is populated by the points whose specializations belong to the level $< i$. The apex of this inverted pyramid is either the *generic point*, (0), if A has no zero divisors, or consists of a finite number of points if A is a Noetherian ring (for a proof, see Sect. 1.4.7b.iii).

Figure 1.5 displays two spectra: the spectrum of the ring of integer p-adic numbers, $\mathbb{Z}_p$, and Spec $\mathbb{C}[T_1, T_2]$. The arrows indicate the specialization relation. The picture of Spec $\mathbb{Z}_p$ does not require comments; note only that Spec A may be a finite, but not a discrete topological space. The other picture is justified by the following statement.

1.4.3a Proposition *Let the field K be algebraically closed. The following list exhausts the prime ideals of the ring $K[T_1, T_2]$:*

(a) *the maximal ideals $(T_1 - t_1, T_2 - t_2)$, where $t_1, t_2 \in K$ are arbitrary;*
(b) *the principal ideals $(F(T_1, T_2))$, where F runs over all irreducible polynomials;*
(c) (0).

For a proof, see [Reid].

The images influenced by this picture can serve as a base for a working dimension theory in algebraic geometry. We will show this later; for the moment we will confine ourselves to preliminary definitions and two simple examples.

We say that a sequence of points $x_0, x_1, \ldots, x_n$ of a topological space X is a *chain of length n with the origin at x_0 and the end at x_n* if $x_i \neq x_{i+1}$ and x_{i+1} is a specialization of x_i for all $0 \leq i \leq n - 1$.

The *height* $\mathrm{ht}(x)$ of $x \in X$ is the upper bound of the lengths of the chains with the origin at x. The *dimension* $\dim X$ of X is the supremum of the heights of its points.

1.4.3a.i Example In the space $X = \operatorname{Spec} K[T_1, \ldots, T_n]$, where K is a field, there is a chain of length n corresponding to the chain of prime ideals

$$(0) \subset (T_1) \subset \cdots \subset (T_1, \ldots, T_n),$$

and therefore $\dim X \geq n$.

Similarly, because there is a chain

$$(p) \subset (p, T_1) \subset \cdots \subset (p, T_1, \ldots, T_n),$$

we conclude that $\dim \operatorname{Spec} \mathbb{Z}[T_1, \ldots, T_n] \geq n + 1$.

As we will see later, in both cases the inequalities are, in fact, equalities.

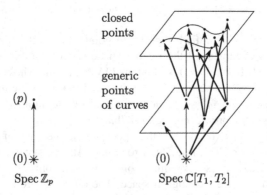

Fig. 1.5

This definition of dimension can be traced back to Euclid: (Closed) points "border" lines, lines "border" surfaces, and so on.[11]

1.4.4 Big Open Sets

For any $f \in A$, we define

$$D(f) = \operatorname{Spec} A \setminus V(f) = \{x \mid f(x) \neq 0\}.$$

The sets $D(f)$ are called *big open sets*; they constitute a basis of the Zariski topology of Spec A, because

$$\operatorname{Spec} A \setminus V(E) = \bigcup_{f \in E} D(f) \quad \text{for any } E \subset A.$$

Consider, for example, Spec $\mathbb{C}[T]$. Its closed points correspond to the ideals $(T - t)$, where $t \in \mathbb{C}$, and thus constitute the "complex line"; the non-empty open sets are $\{0\}$ and the complements of finite sets of points in the complex line. **The closure of any open set coincides with the whole space!**

More generally, if A has no zero divisors and $f \neq 0$, then $D(f)$ is dense in Spec A. Indeed, $D(f) \subset (0)$; hence, $\overline{D(f)} = \overline{(0)} = \operatorname{Spec} A$. Therefore, **any non-empty open set of the spectrum of any ring without zero divisors is dense.**

Analyzing this type of (non-)separability of topological spaces, an important fact was observed:

1.4.5 Lemma *The following conditions are equivalent:*

(a) *Any non-empty open subset of X is dense.*
(b) *Any two non-empty open subsets of X have a non-empty intersection.*
(c) *If $X = X_1 \cup X_2$, where X_1, X_2 are closed, then either $X_1 = X$, or $X_2 = X$.*

Proof (a) $\Longleftrightarrow$ (b) is obvious.

If (c) is false, then there exists a representation $X = X_1 \cup X_2$, where X_1, X_2 are proper closed subsets of X.

Then $X \setminus X_2 = X_1 \setminus (X_1 \cap X_2)$ is a non-dense open set, so (a) does not hold: a contradiction.

Conversely, if (a) does not hold and $U \subset X$ is a non-dense open set, then

$$X = \overline{U} \cup (X \setminus U). \qquad \square$$

---

[11]For a discussion of non-integer values of dimension, with examples, see [Mdim]. As Shander and Palamodov showed, in the super setting, the "bordering" relation can be partly reversed: the *odd* codimension of the boundary of a supermanifold can be *negative*, see [Pa, SoS]. This, not yet rigorously defined concept, corresponds to the fact that the (super)trace of the identity operator on a vector (super)space is (or should be) equal to its (super)dimension.

A topological space X satisfying any of the above conditions is called *irreducible*. Notice that no Hausdorff space with more than one point can be irreducible.

1.4.6 Theorem *Let A be a ring, $\mathfrak{n}(A)$ its nilradical. The space* $\operatorname{Spec} A$ *is irreducible if and only if $\mathfrak{n}(A)$ is prime.*

Proof Let $\mathfrak{n}(A)$ be prime. Since $\mathfrak{n}(A)$ is contained in any prime ideal, it follows that $\operatorname{Spec} A$ is homeomorphic to $\operatorname{Spec} A/\mathfrak{n}(A)$, and $A/\mathfrak{n}(A)$ has no zero divisors.

Conversely, suppose $\mathfrak{n}(A)$ is not prime. It suffices to verify that $\operatorname{Spec} A/\mathfrak{n}(A)$ is reducible, i.e., confine ourselves to the case where A has no nilpotents, but contains zero divisors.

Let $f, g \in A$ be such that $g \neq 0, f \neq 0$, but $fg = 0$. Obviously,

$$\operatorname{Spec} A = V(f) \cup V(g) = V(fg).$$

Therefore, f and g vanish on closed subsets of $\operatorname{Spec} A$, and these closed subsets together cover the whole space $\operatorname{Spec} A$ (this is a natural way in which zero divisors appear in rings of functions).

It remains to verify that $V(f), V(g) \neq \operatorname{Spec} A$. But this is obvious because f and g are not nilpotents. $\qquad\qquad\square$

1.4.6a Corollary *Let $I \subset A$ be an ideal. The closed set $V(I)$ is irreducible if and only if the radical $\mathfrak{r}(I)$ is prime.*

Thus, we obtain a one-to-one correspondence (cf. Corollary 1.4.2a):

> the points of $\operatorname{Spec} A$ $\longleftrightarrow$ the irreducible closed subsets of $\operatorname{Spec} A$

To every point $x \in \operatorname{Spec} A$ there corresponds the closed set $\overline{\{x\}}$; in this situation x is called a *generic point* of this closed set. Every irreducible closed subset obviously has only one generic point.

1.4.7 Decomposition into Irreducible Components

1.4.7a Theorem *Let A be a Noetherian ring. Then* $\operatorname{Spec} A$ *can be uniquely represented as a finite union $\bigcup_i X_i$, where the X_i are maximal closed irreducible subsets.*

Proof The theorem is a geometric reformulation of the ascending chain condition on the ideals of A: every descending chain of closed subsets of $\operatorname{Spec} A$ stabilizes. $\quad\square$

1.4.7b Noetherian Topological Space

Since we will encounter spaces with the property described in Theorem 1.4.7 not homeomorphic to the spectrum of any ring, let us introduce a special definition:

a topological space X will be called *Noetherian* if any descending chain of its closed subsets stabilizes (we say that DCC *holds for X*).

1.4.7b.i Theorem *Let X be a Noetherian topological space. Then X is a finite union of its maximal closed irreducible subsets.*

The maximal closed irreducible subsets of a Noetherian topological space X are called the *irreducible components* of X.

Proof Consider the set of irreducible closed subsets of X, ordered with respect to inclusion. Let us prove that this set is *inductive*, i.e., if $\{X_\alpha \mid \alpha \in J\}$ is a linearly ordered family of irreducible closed subsets of X, then this family has a maximal element for which we may take $\overline{\left(\bigcup_\alpha X_\alpha \right)}$. The irreducibility of the set $\overline{\left(\bigcup_\alpha X_\alpha \right)}$ follows, for example, from the fact that if U_1, U_2 are non-empty open subsets, then $U_1 \cap X_\alpha$ and $U_2 \cap X_\alpha$ are non-empty for some α, and therefore $U_1 \cap U_2$ is non-empty because the X_α are irreducible.

It follows that X is the union of its irreducible components: $X = \bigcup_{i \in I} X_i$.

So far, we have not used the Noetherian property.

Now, let X be Noetherian and $X = X_1 \cup X_2$, where X_1, X_2 are closed. If one of the X_i is reducible, we can represent it as the union of two closed sets, and so on. This process terminates, otherwise we would obtain an infinite descending chain of closed sets (*Noetherian induction*). In the obtained finite union, let us retain only the maximal elements: $X = \bigcup_{1 \leq i \leq n} X_i$. This decomposition coincides with the above one: if Y is an (absolutely) maximal closed subset of X, then $Y \subset \bigcup_{1 \leq i \leq n} X_i$ implies $Y = \bigcup_{1 \leq i \leq n} (X_i \cap Y)$, and therefore $X_i \cap Y = Y$ for some i; hence, $Y = X_i$.

If I' is a proper subset of I, then $\bigcup_{i \in I'} X_i$ no longer coincides with X. Indeed, let X_j be a discarded component, i.e., $j \notin I'$. If $X_j \subset \bigcup_{i \in I'} X_i$, then $X_j = \bigcup_{i \in I'} (X_i \cap X_j)$ and, due to the irreducibility of X_j, we would have $X_i \cap X_j = X_j$ for some $i \in I'$; a contradiction.   □

1.4.7b.ii Corollary *For any Noetherian ring A, the number of its minimal prime ideals is finite.*

Proof Indeed, the minimal prime ideals of Spec A are the generic points of maximal closed subsets, i.e., the irreducible components of Spec A.   □

1.4.7b.iii Corollary *Let A be a Noetherian ring. If all points of Spec A are closed, then the space Spec A is finite and discrete.*

The rings A with this property (all points of Spec A are closed) are called *Artinian*. We recall a common definition: *a ring is Artinian if it satisfies the DCC on ideals*. By a theorem in [ZS], DCC *on ideals implies that every prime ideal is maximal*, so we arrive at another formulation:

Any Artinian ring is a Noetherian ring all prime ideals of which are maximal.

The spectra of Artinian rings resemble very much finite sets in the usual topology. As noted in Sect. 1.3.3 (and in what follows), every point of the spectrum of an Artinian ring is additionally endowed with a multiplicity.

1.4.8 An Interpretation of Zero Divisors

The following theorem will be refined later.

1.4.8a Theorem

(1) *Any element $f \in A$ that vanishes (when considered as a function) on one of the irreducible components of $\operatorname{Spec} A$ is a zero divisor of A.*
(2) *Conversely, if $f \pmod{\mathfrak{n}(A)}$ is a zero divisor in $A/\mathfrak{n}(A)$, where $\mathfrak{n}(A)$ is the nilradical of A, then f vanishes on one of irreducible components of $\operatorname{Spec} A$.*

1.4.8a.i Remark The nilpotents cannot be excluded from item (2) of the theorem: if f is a zero divisor in A, but not in $A/\mathfrak{n}(A)$, then it is possible that f will not vanish on an irreducible component. Here is an example: let A and $B \oplus \mathfrak{a}$, where B is a subring of A without zero divisors, be isomorphic as groups, and $\mathfrak{a} \subset A$ be an ideal with the zero multiplication (all products are equal to 0). Let $\mathfrak{a} \cong B/\mathfrak{p}$ as B-modules, where $\mathfrak{p} \subset B$ is a non-zero prime ideal. Then the elements of $\mathfrak{p}$ are zero divisors in A, because they are annihilated under multiplication by the elements of $\mathfrak{a}$.

On the other hand, clearly, $\operatorname{Spec} A \cong \operatorname{Spec} B$, these spectra are irreducible, and the nonzero elements of $\mathfrak{p}$ cannot vanish on the whole $\operatorname{Spec} A$.

Proof of Theorem Let $\operatorname{Spec} A = X \cup Y$, where X is an irreducible component on which $f \in A$ vanishes, and Y is the union of the other irreducible components. Since Y is closed and $X \not\subset Y$, it follows that there exists an element $g \in A$ that vanishes on Y, but does not vanish identically when restricted to X. Then fg vanishes at all the points of $\operatorname{Spec} A$, so $(fg)^n = 0$ for some n. Hence, $f(f^{n-1}g^n) = 0$. This does not prove yet that f is a zero divisor because it might happen that $f^{n-1}g^n = 0$; but then we may again separate f and continue in this way until we obtain $f^m g^n = 0$ and $f^{m-1}g^n \neq 0$. This will always be the case eventually because $g^n \neq 0$; otherwise, g would have also vanished on X.

Now, let $\bar{f} = f \pmod{N}$ be a zero divisor in N/M, i.e., $\bar{f}\,\bar{g} = 0$ for some $\bar{g} = g$ $\pmod{N}$. Then $\operatorname{Spec} A = \operatorname{Spec} A/N = V(f) \cup V(g)$. Decomposing $V(f)$ and $V(g)$ into irreducible components we see that at least one of the irreducible components of $V(f)$ is also irreducible in $\operatorname{Spec} A$. Otherwise, all irreducible components of $\operatorname{Spec} A$ would be contained in $V(g)$, contradicting the fact that $\bar{g} \neq 0$, i.e., $g \notin N$. Therefore, f vanishes on one of the irreducible components of $\operatorname{Spec} A$, as required.

$\square$

1.4.8b Examples
(1) Let A be a unique factorization ring, $f \in A$. The space $\operatorname{Spec} A/(f) \cong V(f)$ is irreducible if and only if $f = ep^n$, where p is an indecomposable element and e is invertible. This follows directly from Theorem 1.4.6.

In particular, let $A = K[T_1, \ldots, T_n]$, where K is a field. Then $V(f)$ corresponds to the hypersurface (in the affine space) singled out by one equation: $f = 0$. We have just obtained a natural criterion for irreducibility of such a hypersurface.

(2) Let K be a field, $\operatorname{Char} K \neq 2$, and $f \in K[T_1, \ldots, T_n]$ a quadratic form. The equation $f = 0$ determines a reducible set if and only if $\operatorname{rank} f = 2$. Indeed,

reducibility is equivalent to the fact that $f = l_1 \cdot l_2$, where l_1 and l_2 are non-proportional linear forms.

1.4.9 Connected Spaces

The definition of connectedness from general topology is quite suitable for us: a space X is said to be *connected* if it *cannot* be represented as the union of two disjoint non-empty closed subsets. Clearly, any irreducible space is connected.

Any space X can be uniquely decomposed into the union of its maximal connected subspaces, which are pairwise disjoint (for a proof, see [K]) and are called the *connected components* of the space X. Every irreducible component of the space X belongs entirely to one of its connected components. Theorem 1.4.7b implies, in particular, that the number of connected components of a Noetherian space is finite.

The space Spec A is not necessarily connected. For the spectral, as well as for the usual (Hausdorff) topology, we clearly have:

The ring of continuous functions on the disjoint union $X_1 \coprod X_2$ naturally decomposes into the direct product of the rings of functions on X_1 and on X_2. (1.6)

1.4.10 The Decomposition of Spec A Corresponding to a Factorization of A

Let $A_1, \ldots, A_n$ be rings, and let the product $\prod_{1 \leq i \leq n} A_i = A$ be endowed with the structure of a ring with the coordinate-wise addition and multiplication. The set

$$\mathfrak{a}_i = \{x \in A \mid \text{ all coordinates of } x \text{ except the } i\text{-th one are } 0\}$$

is an ideal of A, and $\mathfrak{a}_i \mathfrak{a}_j = 0$ for $i \neq j$. Set

$$\mathfrak{b}_i = \sum_{r \neq i} \mathfrak{a}_r \quad \text{and} \quad X_i = V(\mathfrak{b}_i) \subset \operatorname{Spec} A.$$

Then

$$\bigcup_{1 \leq i \leq n} X_i = V(\mathfrak{b}_1 \cdots \mathfrak{b}_n) = V(0) = \operatorname{Spec} A,$$

$$X_i \cap X_j = V(\mathfrak{b}_i \cup \mathfrak{b}_j) = V(A) = \emptyset \text{ if } i \neq j.$$

Therefore

Spec $\prod_{1\le i\le n} A_i$ splits into the disjoint union of its closed subsets

$$V(\mathfrak{b}_i) \cong \operatorname{Spec} A/\mathfrak{b}_i = \operatorname{Spec} A_i. \tag{1.7}$$

Note that for infinite products the statement (1.7) is false, see Exercise 1.4.14 7(e).

1.4.11   The Statement Converse to (1.6)

1.4.11a Proposition *Let* $X = \operatorname{Spec} A = \coprod_{1\le i\le n} X_i$, *where the* X_i *are closed disjoint subsets. Then there exists an isomorphism* $A \cong \prod_{1\le i\le n} A_i$ *such that, with the notation of the above section,* $X_i = V(\mathfrak{b}_i)$.

Proof We consider the case $n = 2$ in detail. Let $X_i = V(\mathfrak{b}_i)$. By Corollary 1.4.2a, we have

$$X_1 \cup X_2 = X \iff V(\mathfrak{b}_1\mathfrak{b}_2) = X \iff \mathfrak{b}_1\mathfrak{b}_2 \subset N,$$

$$X_1 \cap X_2 = \emptyset \iff V(\mathfrak{b}_1 + \mathfrak{b}_2) = \emptyset \iff \mathfrak{b}_1 + \mathfrak{b}_2 = A.$$

Therefore, there exist elements $f_i \in \mathfrak{b}_i$ and an integer $r > 0$ such that $f_1 + f_2 = 1$ and $(f_1 f_2)^r = 0$.

1.4.11b Lemma *Let* $f_1, \ldots, f_n \in A$ *be such that* $(f_1, \ldots, f_n) = A$.
   Then $(f_1^{m_1}, \ldots, f_n^{m_n}) = A$ *for any positive integers* $m_1, \ldots, m_n$.

Proof By Sect. 1.4, $(g_1, \ldots, g_n) = A$ if and only if $\bigcap_i V(g_i) = \emptyset$. Since $V(g^m) = V(g)$ for $m > 0$, the desired conclusion follows.   □

Thanks to Lemma 1.4.11b, $g_1 f_1^r + g_2 f_2^r = 1$ for some $g_i \in A$. Set $e_i := g_i f_i^r$ for $i = 1, 2$. Then $e_1 + e_2 = 1$ and $e_1 e_2 = 0$, and therefore the $e_i \in \mathfrak{b}_i$ are orthogonal idempotents which determine a factorization of A:

$$A \xrightarrow{\sim} A_1 \times A_2, \quad g \longmapsto (ge_1, ge_2). \tag{1.8}$$

It remains to show that $V(Ae_i) = X_i$. But, clearly, the sets $V(Ae_i)$ do not intersect, their union is X, and $Ae_i \subset \mathfrak{b}_i$; hence, $X_i \subset V(Ae_i)$. Proposition 1.4.11a is proved for $n = 2$.

1.4.11c Exercise Complete the proof by induction on n.   □

1.4.12 Example

Let A be an Artinian ring. Since Spec A is the union of a finite number of closed points, it follows that A is isomorphic to the product of a finite number of local Artinian rings. In particular, any Artinian ring is of finite length (cf. Example 1.3.3).

1.4.13 Quasi-compactness

The usual term "compact" is prefixed with a *quasi-* to indicate that we are speaking about non-Hausdorff spaces. A topological space is said to be *quasi-compact* if every of its open coverings contains a finite subcovering.

The following simple result is somewhat unexpected, because it does not impose any finiteness conditions on A:

1.4.13a Proposition *The space* Spec A *is quasi-compact for any A.*

Proof Any cover of Spec A can be refined to a cover by big open sets: Spec $A = \bigcup_{i \in I} D(f_i)$. Then $\bigcap_i V(f_i) = \emptyset$, so that $(f_i)_{i \in I} = A$. Therefore, there exists a partition of unity $1 = \sum_{i \in I} g_i f_i$, where $g_i \neq 0$ for a finite number of indices $i \in J \subset I$. Thus, Spec $A = \bigcup_{i \in J} D(f_i)$, as desired. □

1.4.14 Exercises

(1) Let $S \subset A$ be a *multiplicative* set.[12] The set S is said to be *complete*, if $fg \in S$ implies $f \in S$ and $g \in S$. Every multiplicative set S has a uniquely determined completion $\widetilde{S}$, namely, the minimal complete multiplicative set containing S.

Show that $D(f) = D(g) \iff \widetilde{(f^n)}_{n \geq 0} = \widetilde{(g^n)}_{n \geq 0}$.

(2) Show that the spaces $D(f)$ are quasi-compact.

(3) Are the following spaces connected?

(3a) Spec $K[T]/(T^2 - 1)$, where K is a field;
(3b) Spec $\mathbb{Z}[T]/(T^2 - 1)$.

(4) The irreducible components of each of the curves

$$T_1(T_1 - T_2^2) = 0 \text{ and } T_2(T_1 - T_2^2) = 0$$

in Spec $\mathbb{C}[T_1, T_2]$ are a straight line and a parabola, and hence are pairwise isomorphic. In both cases the intersection point is the "vertex" of the parabola. Prove that, nevertheless, the rings of functions of these curves are not isomorphic.

---

[12] That is, $1 \in S$ and $f, g \in S \implies fg \in S$.

(5) Let A be a Noetherian ring. Construct a graph whose vertices are in one-to-one correspondence with the irreducible components of the space $\operatorname{Spec} A$, and any two distinct vertices are connected by an edge if the corresponding components have a non-empty intersection. Prove that the connected components of $\operatorname{Spec} A$ are in a one-to-one correspondence with the linearly connected components of the graph.

(6) Finish the proof of Proposition 1.4.11a. Is the decomposition $A = \prod_{1 \le i \le n} A_i$, which is claimed, uniquely determined?

(7) Let $(K_i)_{i \in I}$ be a family of fields. Set $A = \prod_{i \in I} K_i$ and let $\pi_i : A \longrightarrow K_i$ be the projection homomorphisms.

   (a) Let $\mathfrak{a} \subset A$ be a proper ideal. Associate with it a family $\Phi_\mathfrak{a}$ of subsets of I by setting

   $$L \in \Phi_\mathfrak{a} \iff \text{there exists } f \in \mathfrak{a} \text{ such that } \pi_i(f) = 0,$$

   $$\text{if and only if } i \in L.$$

   Show that the subsets L are non-empty and the family $\Phi_\mathfrak{a}$ possesses the following two properties:

   $$(\alpha) \; L_1 \in \Phi_\mathfrak{a} \; \& \; L_2 \in \Phi_\mathfrak{a} \implies L_1 \cap L_2 \in \Phi_\mathfrak{a},$$
   $$(\beta) \; L_1 \in \Phi_\mathfrak{a} \; \& \; L_2 \supset L_1 \implies L_2 \in \Phi_\mathfrak{a}.$$

   (b) The family Φ of non-empty subsets of a set I with the properties (α) and (β) is called a *filter* on the set I.
   Let Φ be a filter on I; assign to it the set $\mathfrak{a}_\Phi \subset A$ by setting

   $$f \in \mathfrak{a}_\Phi \iff \{i \mid \pi_i(f) = 0\} \in \Phi.$$

   Show that $\mathfrak{a}_\Phi$ is an ideal in A.

   (c) Show that the maps $\mathfrak{a} \longmapsto \Phi_\mathfrak{a}$ and $\Phi \longmapsto \mathfrak{a}_\Phi$ establish a one-to-one correspondence between the ideals of A and the filters on I. Further, we see that $\mathfrak{a}_1 \subset \mathfrak{a}_2 \iff \Phi_{\mathfrak{a}_1} \subset \Phi_{\mathfrak{a}_2}$. In particular, to the maximal ideals there correspond maximal filters; they are called *ultrafilters*.

   (d) Let $i \in I$, and $\Phi^{(i)} = \{L \subset I \mid i \in L\}$. Show that $\Phi^{(i)}$ is an ultrafilter. Show that if I is a finite set, then any ultrafilter is of the form $\Phi^{(i)}$ for some i. Which ideals in A correspond to the ultrafilters $\Phi^{(i)}$? What are the quotients of A by these ideals?

   (e) Show that if the set I is infinite, then there exists an ultrafilter on I distinct from $\Phi^{(i)}$.

   Hint Let $\Phi = \{L \subset I \mid I \setminus L\}$ be finite; let $\overline{\Phi}$ be a maximal filter containing Φ. Verify that $\overline{\Phi} \ne \Phi^{(i)}$ for all $i \in I$.

(f) Let $A = \prod_{q \in I} \mathbb{Z}/q\mathbb{Z}$, where I is the set of all primes. Let $\mathfrak{p} \subset A$ be the prime ideal corresponding to an ultrafilter distinct from all $\Phi^{(q)}$. Show that $A/\mathfrak{p}$ is a field of characteristic 0.

1.5 Affine Schemes (A Preliminary Definition)

To any map of sets $f: X \longrightarrow Y$ there corresponds a homomorphism $f^*: F(Y) \longrightarrow F(X)$ of the rings of functions on these sets given by the formula

$$f^*(\varphi)(x) = \varphi(f(x)).$$

If X, Y are *topological spaces* and $F(X)$, $F(Y)$ are rings not of arbitrary, but of *continuous* functions, then the homomorphism f^* is uniquely recovered from the map f whenever f is *continuous*. Without certain conditions this correspondence does not necessarily take place, e.g., the homomorphism f^* of rings of functions of certain class (say, of continuous functions) is **not** uniquely recovered from an arbitrary f.

The main objects of **our** study are the rings of "functions"; therefore, the important maps of spaces are—for us—only those derived from ring homomorphisms.

Let $\varphi: A \longrightarrow B$ be a ring homomorphism. To every prime ideal $\mathfrak{p} \subset B$, we assign its preimage $\varphi^{-1}(\mathfrak{p})$. The ideal $\varphi^{-1}(\mathfrak{p})$ is prime because φ induces the embedding $A/\varphi^{-1}(\mathfrak{p}) \hookrightarrow B/\mathfrak{p}$, and since $B/\mathfrak{p}$ has no zero divisors, neither has $A/\varphi^{-1}(\mathfrak{p})$. We have thus defined a map $^a\varphi: \operatorname{Spec} B \longrightarrow \operatorname{Spec} A$, where the superscript a stands for "affine".

1.5.1 Theorem

(1) $^a\varphi$ *is continuous as a map of topological spaces with respect to the Zariski topologies of these spaces.*
(2) $^a(\varphi\psi) = {}^a\psi \circ {}^a\varphi.$

Proof (2) is obvious. To prove (1), it suffices to verify that the preimage of any closed set is closed. Indeed,

$$y \in V(\varphi(E)) \iff \varphi(E) \subset \mathfrak{p}_y \iff E \subset \varphi^{-1}(\mathfrak{p}_y) = \mathfrak{p}_{^a\varphi(y)}$$

$$\iff {}^a\varphi(y) \in V(E) \iff y \in ({}^a\varphi)^{-1}(V(E)).$$

Therefore, $({}^a\varphi)^{-1}(V(E)) = V(\varphi(E)).$ □

Thus, running a bit ahead, cf. Sect. 1.16, we see that

$$\operatorname{Spec} : \operatorname{Rings}^\circ \longrightarrow \operatorname{Top} \text{ is a functor.}$$

1.5.2 Affine Schemes as Spaces

An *affine scheme* is a triple (X, α, A), where X is a topological space, A a ring, and $\alpha: X \xrightarrow{\sim} \operatorname{Spec} A$ an isomorphism of spaces.

A *morphism of affine schemes* $(Y, \beta, B) \longrightarrow (X, \alpha, A)$ is a pair (f, φ), where $f: Y \longrightarrow X$ is a continuous map and $\varphi: A \longrightarrow B$ is a ring homomorphism, such that the diagram

$$
\begin{array}{ccc}
Y & \xrightarrow{\ \beta\ } & \operatorname{Spec} B \\
\downarrow{\scriptstyle f} & & \downarrow{\scriptstyle {}^a\varphi} \\
X & \xrightarrow{\ \alpha\ } & \operatorname{Spec} A
\end{array}
$$

commutes. The composition of morphisms is defined in an obvious way.

To every ring A, there corresponds the affine scheme $(\operatorname{Spec} A, \operatorname{id}, A)$, where id is the identity map; for brevity, we will often shorten the notation of this affine scheme to $\operatorname{Spec} A$. Clearly, any affine scheme is isomorphic to such a scheme. Affine schemes constitute a category; the dual category is equivalent to the category of rings.

Our definition is not final because it is ill-adjusted to globalization, the gluing of general schemes from affine ones. In what follows we will modify it: an additional element of the structure making the *space* $\operatorname{Spec} A$ into the *scheme* $\operatorname{Spec} A$ is a *sheaf* associated with the ring A. But the above definition will do for a while, because

>   the ring A and the corresponding sheaf of functions
>   on $\operatorname{Spec} A$ can be uniquely recovered from each other.

1.5.3 Examples

To appreciate the difference between the set $\operatorname{Hom}(A, B)$, which is the only set of importance for us here, and the set of all continuous maps $\operatorname{Spec} B \longrightarrow \operatorname{Spec} A$, consider several simple examples.

1.5.3a The Case $A = B = \mathbb{Z}$

As we have already observed, $\operatorname{Spec} \mathbb{Z}$ consists of (0), and the closed points (p), where p runs over the primes. The closure of (0) is the whole space; the remaining closed sets consist of a finite number of closed points. There are lots of automorphisms of the space $\operatorname{Spec} \mathbb{Z}$: we may permute the closed points at random; by contrast, $\operatorname{Hom}(\mathbb{Z}, \mathbb{Z})$ contains only the identity map.

1.5.3b The Case $A = K[T]$, where K is a Finite Field, $B = \mathbb{Z}$

Obviously, Spec A and Spec B are isomorphic as topological spaces, whereas $\mathrm{Hom}(A, B) = \emptyset$.

Examples 1.5.3a and 1.5.3b might make one think that there are considerably fewer homomorphisms of rings than there are continuous maps of their spectra. The opposite phenomenon is, however, also possible.

1.5.3c The Case K is a Field

Then Spec K consists of a single point, and therefore the set of its automorphisms consists of a single point, the identity map; by contrast the group of automorphisms of K may be even infinite (a Lie group). Therefore

> **one-point schemes may have "inner degrees of freedom",**
> **like elementary particles.**

1.5.3d The Spectrum That Embodies an "Idea" of the Vector ("Combing Nilpotents")

Let A be a ring, $B = A[T]/(T^2)$. The natural homomorphism

$$\varepsilon: B \longrightarrow A, \quad a + bt \longmapsto a, \text{ where } t = T \pmod{(T^2)},$$

induces an isomorphism of *topological spaces* $^a\varepsilon: \mathrm{Spec}\, A \longrightarrow \mathrm{Spec}\, B$, but by no means an isomorphism of *schemes*.

The scheme $(\mathrm{Spec}\, B, \mathrm{id}, B)$ is "richer" than $(\mathrm{Spec}\, A, \mathrm{id}, A)$, as it includes the nilpotents tA. To see how this richness manifests itself, consider arbitrary "projections" $^a\pi: \mathrm{Spec}\, B \longrightarrow \mathrm{Spec}\, A$, i.e., scheme morphisms corresponding to the ring homomorphisms $\pi: A \longrightarrow B$ such that $\varepsilon\pi = \mathrm{id}$. Then $\pi(f) - f \in At$. For any such π, define the map $\partial_\pi: A \longrightarrow A$ by setting

$$\pi(f) - f = \partial_\pi(f)t.$$

Since π is a ring homomorphism, we see that $\partial_\pi(f)$ is a derivation of A, i.e., is linear and satisfies the Leibniz rule:

$$\partial_\pi(f + g) = \partial_\pi(f) + \partial_\pi(g),$$
$$\partial_\pi(fg) = \partial_\pi(f) \cdot g + f \cdot \partial_\pi(g).$$

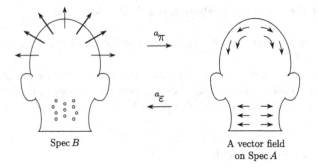

Spec B A vector field
on Spec A

Fig. 1.6 Combing nilpotents

Indeed, the linearity is obvious, while the Leibniz rule follows from the identity

$$\pi(fg) = fg + \partial_\pi(fg)t = \pi(f)\pi(g) = (f + \partial_\pi(f)t)(g + \partial_\pi(g)t),$$

true thanks to the property $t^2 = 0$.

It is easy to see that, conversely, for any derivation $\partial\colon A \longrightarrow A$ the map $\pi\colon A \longrightarrow B$ defined by the formula $\pi(f) = f + (\partial f)t$ is a ring homomorphism and determines a projection $^a\pi$ (Fig. 1.6).

In Differential Geometry, every derivation of the ring of functions on a manifold is interpreted as a "vector field" on the manifold. One can visualize the scheme $(\operatorname{Spec} B, \operatorname{id})$ as being endowed with a field of "sticking out" vectors, as on a hedgehog, as compared with the scheme $(\operatorname{Spec} A, \operatorname{id})$. The morphism $^a\pi$ "combs" these vectors, turning them into a vector field on Spec A.

In particular, if K is a field, then Spec K is a point and

the scheme $(\operatorname{Spec} K[T]/(T^2), \operatorname{id})$ embodies an "idea of the vector"

with the point Spec K as the source of the vector.

In what follows we will sometimes represent nilpotents graphically as arrows, though it is obvious that even for the simplest schemes, like

$$(\operatorname{Spec} K[T_1, T_2]/(T_1^2, T_1 T_2, T_2^2), \operatorname{id}),$$

$$(\operatorname{Spec} K[T]/(T^n), \operatorname{id}) \ \text{ for } n \geq 3,$$

$$\operatorname{Spec}(\mathbb{Z}/(p^2), \operatorname{id}) \ \text{ for } p \text{ prime},$$

such a representation is of limited informational value.

1.5.3e "Looseness" of Affine Spaces

Let K be a field (for the sake of simplicity), V a linear space over K, and $A := S_K^*(V)$ the symmetric algebra of V. Consider the group G of automorphisms of the K-scheme $(\operatorname{Spec} A, \mathrm{id})$. This group consists of transformations induced by those that constitute the group of K-automorphisms of the ring $K[T_1, \ldots, T_n]$, where $n = \dim V$. The group G contains as a subgroup the usual Lie group G_0 of invertible affine transformations

$$T_i \longmapsto \sum_j c_i^j T_j + d_i, \quad \text{where } c_i^j, d_i \in K.$$

For $n = 1$, it is easy to see that $G_0 = G$. This is far from being the case for $n \geq 2$. Indeed, in this case any "triangular" substitution of the form

$$T_1 \longmapsto T_1 + F_1,$$

$$T_2 \longmapsto T_2 + F_2(T_1),$$

$$\cdots\cdots\cdots\cdots\cdots$$

$$T_i \longmapsto T_i + F_i(T_1, \ldots, T_{i-1}),$$

where $F_i \in K[T_1, \ldots, T_{i-1}] \subset K[T_1, \ldots, T_n]$, clearly belongs to G. Therefore, the group of automorphisms of the scheme corresponding to the affine space of dimension ≥ 2 contains non-linear substitutions of arbitrarily high degree. Their existence is used in the proof of the following *Noether normalization theorem.*

1.5.3f Theorem (Noether's Normalization Theorem, [Lang]) *Let $k[x]$, where $x = (x_1, \ldots, x_n)$, be a finitely generated entire ring (aka integral domain) over a field k, and the transcendence degree[13] of $k(x)$ be r. Then there exist elements $y_1, \ldots, y_r$ in $k[x]$ such that $k[x]$ is integral over $k[y]$.*

1.5.3f.i Remark For $n = 2$, the group G is generated by linear and triangular substitutions (Engel, Shafarevich); for $n \geq 3$, this is no longer the case. (*Nagata's conjecture on automorphisms* proved by I. Shestakov and U. Umirbaev.[14]) The explicit description of G is an open (and very tough) problem.[15]

---

[13]Recall that the *transcendence degree* of a field extension L/K is defined as the largest cardinality of an algebraically independent subset of L over K. If no field K is specified, the transcendence degree of a field L is its degree relative to the prime field of the same characteristic, i.e., $\mathbb{Q}$ if L is of characteristic 0, and $\mathbb{F}_p$ if L is of characteristic p.

[14]For a generalization of the result of Shestakov and Umirbaev, see [Ku].

[15]This problem is close to the famous and still open *Jacobian conjecture*, see [EM].

1.5.3g   Linear Projections

Let $V_1 \subset V_2$ be two linear spaces over a field K, and let $X_i = \operatorname{Spec} S_K^{\cdot}(V_i)$. The morphism $X_2 \longrightarrow X_1$ induced by the embedding $S_K^{\cdot}(V_1) \hookrightarrow S_K^{\cdot}(V_2)$ is called the *projection of the scheme X_2 to X_1*. On the sets of K-points it induces the natural map $X_2(K) = V_2^* \longrightarrow V_1^* = X_1(K)$ which restricts every linear functional on V_2 to V_1.

1.6   Topological Properties of Certain Morphisms and the Maximal Spectrum

In this section we study the most elementary properties of the morphisms

$$^a\varphi \colon \operatorname{Spec} B \longrightarrow \operatorname{Spec} A, \qquad \mathfrak{p}_x \longmapsto \varphi^{-1}(\mathfrak{p}_x).$$

This study gives a partial answer to the question: *What is the structure of the topological space $^a\varphi(\operatorname{Spec} B)$?*

As is known (see, e.g., [Lang]), any homomorphism $\varphi \colon A \longrightarrow B$ factorizes into the product of the surjective ring homomorphism $A \longrightarrow A/\operatorname{Ker}\varphi$ and an embedding $A/\operatorname{Ker}\varphi \longrightarrow B$. Let us exhibit properties of $^a\varphi$ in these two cases.

1.6.1   *Properties of $^a\varphi$*

The case where the ring homomorphism $A \longrightarrow A/\operatorname{Ker}\varphi$ is surjective is very simple.

1.6.1a Proposition *Let $\varphi \colon A \longrightarrow B$ be a ring epimorphism. Then $^a\varphi$ is a homeomorphism of $\operatorname{Spec} B$ onto the closed subset $V(\operatorname{Ker}\varphi) \subset \operatorname{Spec} A$.*

1.6.2b Exercise Verify that this is a direct consequence of the definitions.

Hint Prove the continuity of the inverse map

$$(\,^a\varphi)^{-1} \colon V(\operatorname{Ker}\varphi) \longrightarrow \operatorname{Spec} B.$$

In particular, let A be a *ring of finite type* over a field $\mathbb{K}$ or over $\mathbb{Z}$, i.e., A is a quotient of either $\mathbb{K}[T_1, \ldots, T_n]$, or $\mathbb{Z}[T_1, \ldots, T_n]$ with $n < \infty$.

The spectrum of the polynomial ring plays the role of an affine space (over $\mathbb{K}$ or $\mathbb{Z}$, respectively, cf. Example 1.2.1). Therefore, **the spectra of rings of finite type correspond to affine varieties** ("arithmetic affine varieties" if considered over $\mathbb{Z}$): they are embedded into finite-dimensional affine spaces.

Thus, *ring epimorphisms* correspond to *embeddings of spaces*.

Ring monomorphisms do not necessarily induce surjective maps of spectra: only the closure of $^a\varphi(\mathrm{Spec}\,B)$ coincides with $\mathrm{Spec}\,A$. This follows from the following slightly more general result.

1.6.2 Proposition *For any ring homomorphism* $\varphi: A \longrightarrow B$ *and any ideal* $\mathfrak{b} \subset B$, *we have*

$$\overline{^a\varphi(V(\mathfrak{b}))} = V(\varphi^{-1}(\mathfrak{b})).$$

In particular, if $\mathrm{Ker}\,\varphi = 0$, *then* $^a\varphi(V(0)) = V(0)$, *i.e., the image of* $\mathrm{Spec}\,B$ *is dense in* $\mathrm{Spec}\,A$.

Proof We may assume that $\mathfrak{b}$ is a radical ideal, because $V(\mathfrak{r}(\mathfrak{b})) = V(\mathfrak{b})$ and $\varphi^{-1}(\mathfrak{r}(\mathfrak{b})) = \mathfrak{r}(\varphi^{-1}(\mathfrak{b}))$. The set $\overline{^a\varphi(V(\mathfrak{b}))}$ is the intersection of all closed subsets containing $^a\varphi(V(\mathfrak{b}))$, i.e., the set of common zeroes of all functions $f \in A$ that vanish on $^a\varphi(V(\mathfrak{b}))$. But f vanishes on $^a\varphi(V(\mathfrak{b}))$ if and only if $\varphi(f)$ vanishes on $V(\mathfrak{b})$, i.e., if and only if $\varphi(f) \in \mathfrak{b}$ (because $\mathfrak{b}$ is a radical ideal) or, finally, if and only if $f \in \varphi^{-1}(\mathfrak{b})$. Therefore, the closure we are interested in is equal to $V(\varphi^{-1}(\mathfrak{b}))$.   □

Now, let us give examples of ring monomorphisms for which $^a\varphi(\mathrm{Spec}\,B)$ does not actually coincide with $\mathrm{Spec}\,A$.

1.6.2a   Examples

(1) *Projection of a hyperbola to a coordinate axis.* For a field K, consider the embedding

$$\varphi: A = K[T_1] \longhookrightarrow K[T_1, T_2]/(T_1 T_2 - 1) = B.$$

Then $^a\varphi(\mathrm{Spec}\,B) = D(T_1)$, in accordance with Fig. 1.7. Indeed, $^a\varphi$ maps a generic point into a generic one. The prime ideal $(f(T_1)) \subset A$, where

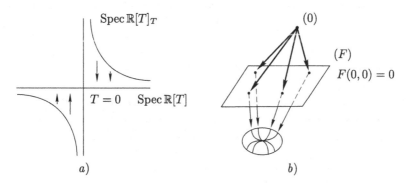

Fig. 1.7

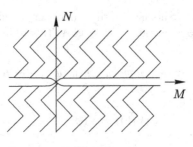

Fig. 1.8

$f \neq cT_1$ is an irreducible polynomial, is the preimage of the prime ideal
$(f(T_1))$ (mod $T_1 T_2 - 1) \subset B$. Finally, T_1 and $T_1 T_2 - 1$ generate the ideal
$(1) = K[T_1, T_2]$, and therefore $(T_1) \notin {}^a\varphi(\operatorname{Spec} B)$.   □

In this example, ${}^a\varphi(\operatorname{Spec} B)$ is open; but it may be neither open, nor closed:

(2) *Projection of a hyperbolic paraboloid onto the plane.* Consider the homomorphism

$$\varphi: A = K[M, N] \longrightarrow B = K[M, N, T]/(MT - N).$$

1.6.2b Exercise Verify that ${}^a\varphi(\operatorname{Spec} B) = D(M) \cup V(M, N)$, and that this set is
indeed not open (it is obvious that it is not closed).

This example, see also Fig. 1.8, illustrates a phenomenon noted long ago in the
study of equations:

The set ${}^a\varphi(\operatorname{Spec} B)$ is the set of values of the coefficients M, N in a K-algebra
for which the equation $MT - N = 0$ is solvable for T.

In general, the solvability condition is the inequality $M \neq 0$, but even for $M = 0$
the solvability is guaranteed if, in addition, $N = 0$.

One can prove that if A is Noetherian and the A-algebra B has a finite number
of generators, then ${}^a\varphi(\operatorname{Spec} B)$ is the union of a finite number of *locally closed*
sets.[16] Such unions are called *constructible* sets; under the described conditions the
image of a constructible set with respect to ${}^a\varphi$ is always constructible (*Chevalley's
theorem* [AM]).

In terms of undetermined coefficients of a (finite) system of equations this means
that the compatibility condition for the system reads:

The coefficients of the system of equations must satisfy one of a finite
number of statements, each statement being a finite collection of polynomial
equalities and inequalities (with zeros as right-hand sides).

---

[16] A *locally closed* set is the intersections of a closed set and an open set.

For example, for the equation $MT - N = 0$, one has the two statements:

(1) $M \neq 0$;
(2) $M = 0, N = 0$.

1.6.3 Analogs of "Finite-sheeted Coverings" of Riemann Surfaces

In the above examples, something "escaped to infinity". Let us describe an important class of morphisms $^a\varphi$ for which this does not happen.

Let B be an A-algebra. An element $x \in B$ is said to be *integral over A* if it satisfies an equation of the form

$$x^n + a_{n-1}x^{n-1} + \cdots + a_0 = 0, \quad a_0, \ldots, a_{n-1} \in A,$$

and B is said to be *integral over A* if all its elements are integral over A.

There are two important cases in which it is easy to establish whether B is integral over A.

Case 1 If B has a finite number of generators as an A-module, then B is integral over A.

Indeed, if A is Noetherian ring, then, for any $g \in B$, the ascending sequence of A-modules

$$B_k = \sum_{0 \leq i \leq k} Ag^i \subset B$$

stabilizes. Therefore, for some k, we have $g^k \in \sum_{0 \leq i \leq k-1} Ag^i$, which yields an integral dependence relation.

The general case reduces to the above one with the help of the following trick. Let $B = \sum_{1 \leq i \leq n} Af_i$. Let

$$f_i f_j = \sum_{1 \leq k \leq n} a_{ij}^k f_k, \text{ where } a_{ij}^k \in A, \text{ and } g = \sum_{1 \leq i \leq n} g_i f_i, \text{ where } g_i \in A;$$

denote by $A_0 \subset A$ the smallest subring containing all a_{ij}^k and g_i; set $B_0 = \sum_{1 \leq i \leq n} A_0 f_i$. Obviously, A_0 is a Noetherian ring, B is an A_0-algebra, and $g \in B_0$. Therefore, g satisfies an integral dependence relation with coefficients in A.

Case 2 Let G be a finite subgroup of the group of automorphisms of B and $A = B^G$ the subring of G-invariant elements. Then B is integral over A.

Indeed, all elementary symmetric polynomials in $s(g)$, where $s \in G$, belong to A for any $g \in B$, and g satisfies $\prod_{s \in G}(g - s(g)) = 0$. $\square$

1.6.4   Localization at a Multiplicative Set

We would like to define the ring of fractions f/g, where g runs over a set S of elements of an arbitrary ring A. When we add or multiply fractions their denominators are multiplied; hence, S should be closed under multiplication. Additionally, we require S to contain the unit element, i.e., to be what is called a *multiplicative set*.

Examples of multiplicative sets S:

(1) $S_f = \{f^n \mid n \in \mathbb{Z}_+\}$ for any $f \in A$.
(2) $S_\mathfrak{p} := A \setminus \mathfrak{p}$ for any prime ideal $\mathfrak{p}$.

The set $S_\mathfrak{p}$ is indeed a multiplicative set because it consists of "functions" $f \in A$ that do not vanish at $\{\mathfrak{p}\} \in \operatorname{Spec} A$; now recall the definition of the prime ideal.

For any multiplicative set S of the ring A, we define the *ring of fractions* A_S (also denoted $S^{-1}A$ or $A[S^{-1}]$), or the *localization of A at S*, as follows. As a set, A_S is the quotient of $A \times S$ modulo the following equivalence relation:

$$(f_1, s_1) \sim (f_2, s_2) \iff$$

there exists $t \in S$ such that $t(f_1 s_2 - f_2 s_1) = 0$.

Denote by f/s the class of the element (f, s), and define the composition law in A_S by the usual formulas:

$$f/s + g/t = (ft + gs)/st, \quad (f/s) \cdot (g/t) = fg/st.$$

1.6.4a Exercise Verify that the above equivalence relation is well defined. The unit of A_S is the element $1/1$, and zero of A_S is the element $0/1$.

Quite often the localizations with respect to S_f and $S_\mathfrak{p}$ are shortly (and somewhat inconsistently) denoted by A_f and $A_\mathfrak{p}$, respectively.

For rings without zero divisors, the map $a \longmapsto a/1$ embeds A into A_S. In general, however, a nontrivial kernel may appear, as described in the following obvious lemma.

1.6.4b Lemma *Let $j: A \longrightarrow A_S$ be the map $a \longmapsto a/1$. Then*

(1) *j is a ring homomorphism and*

$$\operatorname{Ker} j = \{f \in A \mid \text{there exists } s \in S \text{ such that } sf = 0\};$$

(2) *if $0 \notin S$, then all elements of $j(S)$ are invertible in A_S; in the opposite case, $A_S = 0$;*

(3) *every element of A_S can be represented in the form $j(f)/j(s)$ with $s \in S, f \in A$.*

Notice how drastically A_S shrinks when we introduce zero divisors in S.

The following theorem is a basic fact on rings of fractions, it describes the universal character of localization.

1.6.4c Theorem *Let S be a multiplicative set in the ring A and let $j: a \longmapsto a/1$ be the canonical homomorphism $A \longrightarrow A_S$. For any ring homomorphism $f: A \longrightarrow B$ such that every element of $f(S)$ is invertible, there exists a unique homomorphism $f': A_S \longrightarrow B$ such that $f = f' \circ j$.*

1.6.4d Exercise Prove Theorem 1.6.4c.

1.6.4e Corollary *Let A be a ring, and let T and S be multiplicative sets in A such that $T \supset S$. Then*

(1) *the following diagram commutes:*

$$
\begin{array}{ccc}
A & \xrightarrow{\;\; j_S \;\;} & A_S \qquad \ni \; j_S(a)/j_S(t) \\
& \searrow{\scriptstyle j_T} \quad \nearrow{\scriptstyle j_T(S^{-1})} & \Big\downarrow \\
& A_T & \qquad \ni \; j_T(a)/j_T(t)
\end{array}
$$

(2) $A[T^{-1}] \cong A[S^{-1}][j_S(T)^{-1}]$.

1.6.4f Theorem *Let $j: A \longrightarrow A_S$ be the canonical homomorphism. Then the induced map $\;^a j: \operatorname{Spec} A_S \longrightarrow \operatorname{Spec} A$ homeomorphically maps $\operatorname{Spec} A_S$ onto the subset $\{x \in \operatorname{Spec} A \mid \mathfrak{p}_x \cap S = \varnothing\}$.*

Proof We may confine ourselves to the case $0 \notin S$. First, let us prove that there is a one-to-one correspondence

$$\operatorname{Spec} A_S \longleftrightarrow \{x \in \operatorname{Spec} A \mid \mathfrak{p}_x \cap S = \varnothing\}.$$

Let $y \in \operatorname{Spec} A_S$ and $x = \;^a j(y)$. Then $\mathfrak{p}_x \cap S = \varnothing$, because otherwise $\mathfrak{p}_y$ would contain the image under j of an element of S, which is invertible, so $\mathfrak{p}_y$ would contain the unit element.

Now, let $x \in \operatorname{Spec} A$ and $\mathfrak{p}_x \cap S = \varnothing$. Set $\mathfrak{p}_y := \mathfrak{p}_x[S^{-1}]$. Let us show that the ideal $\mathfrak{p}_y$ is prime. Indeed, let $(fs^{-1})(gt^{-1}) \in \mathfrak{p}_y$. Then $fg \in \mathfrak{p}_x$ and, because $\mathfrak{p}_x$ is prime, it follows that either $f \in \mathfrak{p}_x$ or $g \in \mathfrak{p}_x$, and therefore either $fs^{-1} \in \mathfrak{p}_y$ or $gt^{-1} \in \mathfrak{p}_y$.

The fact that the maps $\mathfrak{p}_y \longmapsto j^{-1}(\mathfrak{p}_y)$ and $\mathfrak{p}_x \longmapsto \mathfrak{p}_x[S^{-1}]$ are mutually inverse can be established as follows.

Consider the set $\{f \in A \mid f/1 \in \mathfrak{p}_x[S^{-1}]\}$. Let $f/1 = f'/s$, where $f' \in \mathfrak{p}_x$. Multiplying by some $t \in S$ we obtain $tf \in \mathfrak{p}_x$, whence $f \in \mathfrak{p}_x$.

It remains to show that this one-to-one correspondence is a homeomorphism. As proved earlier, the map $\;^a j$ is continuous. Therefore, it suffices to prove that this map takes every closed set into a closed one, i.e., $j(V(E)) = V(E')$ for some E'. Indeed, take

$$E' = \{\text{denominators of the elements of } E\}. \qquad\qquad \square$$

1.6.4g Example Let us show that the set $D(f)$ is homeomorphic to $\operatorname{Spec} A_f$. Indeed,

$$\mathfrak{p}_x \cap \{f^n \mid n \in \mathbb{Z}\} = \emptyset \Longleftrightarrow f \notin \mathfrak{p}_x.$$

Therefore, $\operatorname{Spec} A$ splits into the union of an open set and a closed set, namely:

$$\operatorname{Spec} A = \operatorname{Spec} A_f \cup \operatorname{Spec} A/(f).$$

Here a certain "duality" between localization and passage to the quotient ring is reflected. Considering $D(f)$ as $\operatorname{Spec} A_f$ we "send $V(f)$ to infinity".

If S is generated by a finite number of elements $f_1, \ldots, f_n$, then

$$\mathfrak{p}_x \cap S = \emptyset \Longleftrightarrow x \in \bigcap_{1 \le i \le n} D(f_i),$$

so in this case the image of $\operatorname{Spec} A_S$ is open in $\operatorname{Spec} A$. However, this is not always the case; cf. Examples 1.6.2.

If $S = A \setminus \mathfrak{p}_x$, then the image of $\operatorname{Spec} A_S$ in $\operatorname{Spec} A$ consists of all points $y \in \operatorname{Spec} A$ whose specialization is x. It is not difficult to see that in general this set is neither open, nor closed in $\operatorname{Spec} A$: *It is the intersection of all the open sets containing x* (Fig. 1.9).

The ring $A_{\mathfrak{p}_x} = \mathcal{O}_x$ contains the unique maximal ideal $\mathfrak{p}_x$. The spectrum of $\mathcal{O}_x$ geometrically describes the "neighborhood of x" in the sense that in a "vicinity" of x

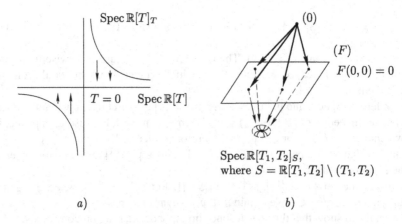

a)   b)

Fig. 1.9

one can track the behavior of all irreducible subsets of Spec A containing x. This is the algebraic version of a *germ of neighborhoods* of x.   □

1.6.5 Theorem *Let* $\varphi: A \hookrightarrow B$ *be a monomorphism and let* B *be integral over* A. *Then* $^a\varphi(\operatorname{Spec} B) = \operatorname{Spec} A$.

Proof We first prove two particular cases and then reduce the general statement to them.

Case 1 B is a field. Then $^a\varphi(\operatorname{Spec} B) = \{(0)\} \subset \operatorname{Spec} A$ and $^a\varphi$ is an epimorphism if and only if A has no other prime ideals, i.e., A is a field. Let us verify this.

Let $f \in A, f \neq 0$. Let us show that if f is invertible in B, then it is so in A, because f^{-1} is integral over A, i.e., satisfies an equation

$$f^{-n} + \sum_{1 \leq i \leq n-1} a_i f^{-i} = 0, \quad a_i \in A.$$

Then, multiplying by f^{n-1}, we obtain

$$f^{-1} = - \sum_{0 \leq i \leq n-1} a_i f^{n-i-1} \in A,$$

as desired.

Case 2 A is a local ring. Then, under the conditions of the theorem, the unique closed point of Spec A belongs to $^a\varphi(\operatorname{Spec} B)$ and, moreover, it is the image under $^a\varphi$ of any other closed point of Spec B.

Indeed, let $\mathfrak{p}$ be a maximal ideal of A, $\mathfrak{q}$ a maximal ideal of B. Then $B/\mathfrak{q}$ is a field integral over the subring $A/(A \cap \mathfrak{q})$ which by Case 1 proved above, must also be a field. This means that $A \cap \mathfrak{q}$ is a maximal ideal in A, hence $A \cap \mathfrak{q} = \varphi^{-1}(\mathfrak{q}) = \mathfrak{p}$.

Case 3: The General Case Let $\mathfrak{p} \subset A$ be a prime ideal; we want to show that there exists an ideal $\mathfrak{q} \subset B$ such that $\mathfrak{q} \cap A = \mathfrak{p}$. Set $S := A \setminus \mathfrak{p}$.

Considering S as a subset of both A and B, we can construct the rings of fractions $A_S \subset B_S$. Set $\mathfrak{p}_S := \{f/s \mid f \in \mathfrak{p}, \ s \in S\}$. It is easy to see that $\mathfrak{p}_S \subset A_S$ is a prime ideal. It is maximal because $A_S \setminus \mathfrak{p}_S$ consists of invertible elements $s/1$.

The ring B_S is integral over A_S: if $f \in B$ satisfies $f^n + \sum_{0 \leq i \leq n-1} a_i f^i = 0$, then $f/s \in B_S$ satisfies

$$(f/s)^n + \sum_{0 \leq i \leq n-1} (a_i/s^{n-i}) \cdot (f/s)^i = 0.$$

Therefore, by Case 2, there is a prime ideal $\mathfrak{q}_S \subset B_S$ such that $A_S \cap \mathfrak{q}_S = \mathfrak{p}_S$.

The preimage of $\mathfrak{q}_S$ in B under the natural homomorphism $B \longrightarrow B_S$ is a prime ideal. It remains to verify that $A \cap \mathfrak{q} = \mathfrak{p}$, The inclusion $\mathfrak{p} \subset A \cap \mathfrak{q}$ is obvious.

Let $f \in A \cap \mathfrak{q}$. There exist $n \in \mathbb{N}$ and $s \in S$ such that $f/s^n \in \mathfrak{q}_S$. Therefore, $f/s^n \in A_S \cap \mathfrak{q}_S = \mathfrak{p}_S$, so that $s^m f \in \mathfrak{p}$ for some $m \geq 0$; hence, $f \in \mathfrak{p}$.   □

In this proof the ring of fractions A_S appeared as a technical device which enabled us to "isolate"—localize—the prime ideal $\mathfrak{p} \subset A$, producing from it a maximal ideal $\mathfrak{p}_S$ in A_S. The term "localization" is applied to the construction of rings of fractions with precisely this geometric meaning.

1.6.6   Addendum to Theorem 1.6.5

Denote by $\operatorname{Spm} A$ the set of *maximal* ideals of A (the "maximal spectrum").

1.6.6a Proposition   *Under the conditions of Theorem 1.6.5 we have*

(1) $^a\varphi(\operatorname{Spm} B) = \operatorname{Spm} A$;
(2) $(^a\varphi)^{-1}(\operatorname{Spm} A) = \operatorname{Spm} B$.

Proof (1) Let $\mathfrak{p} \in \operatorname{Spm} B$; then $B/\mathfrak{p}$ is a field integral over $A/(A \cap \mathfrak{p}) = A/\varphi^{-1}(\mathfrak{p})$. Thanks to Theorem 1.6.5, Case (1), $A/\varphi^{-1}(\mathfrak{p})$ is also a field; therefore, $\varphi^{-1}(\mathfrak{p})$ is maximal in A. This demonstrates that $^a\varphi(\operatorname{Spm} B) = \operatorname{Spm} A$.

To prove item (2), consider a prime ideal $\mathfrak{q} \subset B$ such that $\mathfrak{p} = A \cap \mathfrak{q} \subset A$ is maximal. The ring without zero divisors $B/\mathfrak{q}$ is integral over the field $A/\mathfrak{p}$ and we need to verify that it is also a field. Indeed, any $f \in B/\mathfrak{q}$, being integral over $A/\mathfrak{p}$, belongs to the finite-dimensional $A/\mathfrak{p}$-algebra generated by the powers of f. Multiplication by f in this algebra is a linear map with zero kernel, and therefore an epimorphism. In particular, the equation $fu = 1$ is solvable, proving the statement.   □

1.6.6b   Warning

Let $\varphi: A \longrightarrow B$ be a ring homomorphism, and $x \in \operatorname{Spm} B$, $y \in \operatorname{Spm} A$. In general, the point $^a\varphi(x)$ is nonclosed, and $(^a\varphi)^{-1}(y)$ contains also nonclosed points; so Proposition 1.6.6a describes a rather particular situation.

1.6.6c Example   If $\varphi: \mathbb{Z}_p \longhookrightarrow \mathbb{Z}_p[T]$ is a natural embedding, then $\mathfrak{p}_x = (1 - pT)$ is, clearly, a maximal ideal in $\mathbb{Z}_p[T]$ (the quotient by $\mathfrak{p}_x$ is isomorphic to $\mathbb{Q}_p$). Obviously,

$$\varphi^{-1}(\mathfrak{p}_x) = \mathbb{Z}_p \cap (1 - pT) = (0),$$

and therefore $^a\varphi(x) \notin \operatorname{Spm} \mathbb{Z}_p$, where $x \in \operatorname{Spec} \mathbb{Z}_p[T]$ is the point corresponding to $\mathfrak{p}_x$. Moreover, the image of the closed point x is a generic point of $\operatorname{Spec} \mathbb{Z}_p$; this generic point is an open set, being the complement to (p).

Now let $\mathfrak{p}_y = (p) \subset \mathbb{Z}_p[T]$ and $\mathfrak{p}_x = (p) \subset \mathbb{Z}_p$. Then $y \in (^a\varphi)^{-1}(x)$, and x is closed, whereas y is not.

This, however, is not unexpected. Even more transparent is the example of the projection of the plane onto the straight line given by the following embedding of the corresponding rings:

$$K[T_1] \lhook\joinrel\longrightarrow K[T_1, T_2], \quad T_1 \longmapsto T_1.$$

The preimage of the point $T_1 = 0$ of this line contains, clearly, the generic point of the T_2-axis, which is not closed in the plane. In particular, the correspondence

$$A \longmapsto \operatorname{Spm} A \text{ is not a functor, unlike } A \longmapsto \operatorname{Spec} A.$$

1.6.7 Exercises

(1) Let B be an A-algebra. Prove that the elements of B that are integral over A constitute an A-subalgebra of B.
(2) Let $A \subset B \subset C$ be rings; suppose B is integral over A, and C is integral over B. Prove that C is integral over A.
(3) Let A be a unique factorization ring. Then A is an integrally closed ring in its ring of fractions, i.e., any fraction f/g integral over A belongs to A.

1.7   Closed Subschemes and the Primary Decomposition

In this section we will often denote the subscheme $(V(\mathfrak{a}), \alpha, S)$ by $\operatorname{Spec} A/\mathfrak{a}$ and omit the word "closed" because no other subschemes will be considered here.

1.7.1   Reduced Schemes: Closed Embeddings

Let $X = \operatorname{Spec} A$ be an affine scheme, $\mathfrak{a} \subset A$ an ideal. The scheme $(V(\mathfrak{a}), \alpha, A/\mathfrak{a})$, where $\alpha \colon V(\mathfrak{a}) \xrightarrow{\;\simeq\;} \operatorname{Spec} A/\mathfrak{a}$ is the canonical isomorphism (see Sect. 1.6.1) is said to be a *closed subscheme* of X corresponding to $\mathfrak{a}$. Therefore

> *the closed* **subschemes** *of the* **scheme** $X = \operatorname{Spec} A$ *are*
> *in a one-to-one correspondence with the ideals of* A,

unlike the closed **subsets** of the **space** $\operatorname{Spec} A$, which only correspond to the *radical* ideals.

The *support* of the subscheme $Y = \operatorname{Spec} A/\mathfrak{a} \subset X$ is the space $V(\mathfrak{a})$; it is denoted by $\operatorname{supp} Y$.

To the projection $A \longrightarrow A/\mathfrak{a}$ there corresponds an embedding $Y \longrightarrow X$, called a *closed embedding* of a subscheme.

For any ring L, we introduced the set

$$X(L) := \mathrm{Hom}(\mathrm{Spec}\, L, X) := \mathrm{Hom}(A, L),$$

and called it the *set of L-points* of X (cf. Sect. 1.2.1). Clearly, the L-points of a subscheme Y constitute a subset $Y(L) \subset X(L)$ and the functor $L \longmapsto Y(L)$ is a subfunctor of the functor $L \longmapsto X(L)$.

There is a natural *order* on the set of closed subschemes of a scheme X: we write $Y_1 \subset Y_2$ if $\mathfrak{a}_1 \supset \mathfrak{a}_2$, where $\mathfrak{a}_i$ is the ideal that determines Y_i. The use of the inclusion sign is justified by the fact that

$$Y_1 \subset Y_2 \iff Y_1(L) \subset Y_2(L) \ \text{ for every ring } L.$$

The relation "Y is a closed subscheme of X" is transitive in the obvious sense.

For any closed subset $V(E) \subset X$, there exists a unique minimal closed subscheme with $V(E)$ as its support: it is determined by the ideal $\mathfrak{r}((E))$ and its ring has no nilpotents. Such schemes are said to be *reduced*.

In particular, the subscheme $\mathrm{Spec}\, A/\mathfrak{n}(A)$, where $\mathfrak{n}(A)$ is the nilradical, is the smallest closed subscheme whose support is the whole space $\mathrm{Spec}\, A$. If $X = \mathrm{Spec}\, A$, then its reduction $\mathrm{Spec}\, A/\mathfrak{n}(A)$ is denoted by X_{red}. Thus, a scheme X is reduced if $X = X_{\mathrm{red}}$.

1.7.2   Intersections

The *intersection* $\bigcap_i Y_i$ of a family of subschemes Y_i, where $Y_i = \mathrm{Spec}\, A/\mathfrak{a}_i$, is the subscheme determined by the ideal $\sum_i \mathfrak{a}_i$.

The notation is justified by the fact that, for any ring L, the set of L-points $(\bigcap_i Y_i)(L)$ of the intersection is naturally identified with $\bigcap_i Y_i(L)$. Indeed, an L-point $\varphi: A \longrightarrow L$ belongs to $\bigcap_i Y_i(L)$ if and only if $\mathfrak{a}_i \subset \mathrm{Ker}\, \varphi$ for all i, which is equivalent to the inclusion $\sum_i \mathfrak{a}_i \subset \mathrm{Ker}\, \varphi$. This argument shows also that

$$\mathrm{supp}\left(\bigcap_i Y_i \right) = \bigcap_i \mathrm{supp}\, Y_i.$$

1.7.3   Quasiunions

The notion of the *union of a family of subschemes* is *not* defined in a similar way. In general, for given Y_i's, there is no closed subscheme Y such that $Y(L) = \bigcup_i Y_i(L)$

for all L. However, there exists a smallest subscheme Y such that

$$Y(L) \supset \bigcup_i Y_i(L) \quad \text{for all } L.$$

This Y is determined by the ideal $\bigcap_i \mathfrak{a}_i$. Indeed, if $Y(L) \supset \bigcup_i Y_i(L)$, then the ideal $\mathfrak{a}$ that determines Y satisfies the condition

> "any ideal containing one of the $\mathfrak{a}_i$ contains $\mathfrak{a}$."   (1.9)

The sum $\mathfrak{a}$ of all such ideals also satisfies (1.9) and is the unique maximal element of the set of ideals satisfying (1.9); on the other hand, all elements of this set are contained in $\mathfrak{a}_i$, and therefore in $\bigcap_i \mathfrak{a}_i$.

Having failed to define the union of subschemes, let us introduce a notion corresponding to $\bigcap_i \mathfrak{a}_i$: the *quasiunion* $\bigvee_i Y_i$ of a family of closed subschemes Y_i of the scheme X is the subscheme corresponding to the intersection of all ideals defining the subschemes Y_i.

It is important to notice that the quasiunion of the subschemes Y_i does not depend on the choice of the closed scheme containing all Y_i, inside of which we construct this quasiunion.

The main aim of this section is to construct, for the Noetherian affine schemes, a theory of decomposition into "irreducible" (in some sense) components similar to the one constructed above for Noetherian topological spaces. To this end, we use quasiunions; on supports a quasiunion coincides with the union (for finite families of subschemes).

1.7.3a Lemma $\mathrm{supp}\left(\bigvee_{1 \le i \le n} Y_i \right) = \bigcup_{1 \le i \le n} \mathrm{supp}\, Y_i.$

Proof The inclusion $\supset$ is already proved.

Conversely, if $x \notin \bigcup_{1 \le i \le n} \mathrm{supp}\, Y_i$, then for every i there exists an element $f_i \in \mathfrak{a}_i$ such that $f_i(x) \ne 0$, and therefore $\left(\prod_{1 \le i \le n} f_i \right)(x) \ne 0$. So, x does not belong to the set of zeroes of all functions in the set $\bigcap_{1 \le i \le n} \mathfrak{a}_i$, which is $\mathrm{supp}\left(\bigvee_{1 \le i \le n} Y_i \right)$. To conclude, apply Exercise 1.3.4 (1). $\square$

1.7.4 Irreducible Schemes

Now, we have to transfer the notion of irreducibility to subschemes. The first approach that comes to mind is to try to imitate the definition of irreducibility for spaces.

An affine scheme X is said to be *reducible* if one can write $X = X_1 \vee X_2$, where X_1 and X_2 are proper closed subschemes of X; it is said to be *irreducible* otherwise.

An affine scheme X is said to be *Noetherian* if its ring of global functions is Noetherian or, equivalently, if X satisfies the descending chain condition on closed subschemes.

1.7.1a Theorem *Any Noetherian affine scheme X decomposes into the quasiunion of a finite number of closed irreducible subschemes.*

Proof The same arguments as at the end of Sect. 1.4.7b lead to the result desired.

If X is reducible, we write $X = X_1 \vee X_2$ and then decompose, if necessary, X_1 and X_2, and so on. Thanks to the Noetherian property, the process terminates.   □

1.7.5 Primary Schemes and Primary Ideals

The above notion of irreducibility turns out to be too subtle. The following notion of primary affine schemes is more useful: an ideal $\mathfrak{q} \subset A$ is called *primary* if any zero divisor in $A/\mathfrak{q}$ is nilpotent. A *closed subscheme* is called *primary* if it is determined by a primary ideal.

1.7.5a Proposition *Any irreducible Noetherian scheme is primary.*

1.7.5b Remark The converse statement is false. Indeed, let K be an infinite field. Consider the ring $A = K \times V$, where V is an ideal with the zero multiplication. The ideal (0) is primary and, for any subspace $V' \subset V$, the ideal $(0, V')$ is primary. At the same time, if $\dim_K V > 1$, then there exist infinitely many representations of the form $(0) = V_1 \cap V_2$, where $V_i \subset V$ are proper subspaces, i.e., representations of the form $X = Y_1 \vee Y_2$, where $X = \operatorname{Spec} A$. This deprives us of any hope for uniqueness of the decomposition into a quasiunion of irreducible subschemes. Primary subschemes behave a sight more nicely than irreducible subschemes, as we will see shortly.

Proof of Proposition 1.7.5a Let us show that every non-primary Noetherian scheme X is reducible. Indeed, in the ring of its global functions A there are two elements f, g such that $fg = 0$, $g \neq 0$, and f is not nilpotent.

Set $\mathfrak{a}_k = \operatorname{Ann} f^k := \{h \in A \mid hf^k = 0\}$. The set of ideals $\mathfrak{a}_k$ ascends, and therefore stabilizes. Let n be the smallest number such that $\mathfrak{a}_n = \mathfrak{a}_{n+1}$.

Then $(0) = (f^n) \cap (g)$.

Indeed, $h \in (f^n) \cap (g)$; hence, $h = h_1 f^n = h_2 g$; but $h_1 f^{n+1} = h_2 g f = 0$, and so $h_1 f^n = 0$, because $\mathfrak{a}_{n+1} = \mathfrak{a}_n$. Therefore, $Y = Y_1 \vee Y_2$, where Y_1 is determined by (f^n) and Y_2 by (g).   □

1.7.6 Remarks
(a) The support of any primary Noetherian scheme is irreducible. Indeed, the radical of any primary ideal is a prime ideal.

(b) The results of Sects. 1.7.4 and 1.7.5 show that any affine Noetherian scheme is a quasi-union of its primary subschemes $X = \bigvee_{1 \leq i \leq n} X_i$. We could have kept in this decomposition only maximal elements and then try to prove its uniqueness in the same way as we have done for spaces at the end of Sect. 1.4.7b. But these arguments fail twice: first, the formula

$$X \cap \left(\bigvee_{1 \leq i \leq n} Y_i \right) = \bigvee_{1 \leq i \leq n} (X \cap Y_i)$$

is false in general, and, second, as we have established, our primary subschemes X_i can themselves be reducible.

Therefore, instead of discarding non-maximal components of $\bigvee_{1 \leq i \leq n} Y_i$, we should apply a less trivial process, and even after that the uniqueness theorem will be hard to formulate and prove.

1.7.7 Incontractible Primary Decompositions

A primary decomposition $X = \bigvee_{1 \leq i \leq n} X_i$ is called *incontractible* if the following two conditions are satisfied:

(1) $\operatorname{supp} X_i \neq \operatorname{supp} X_j$ if $i \neq j$,
(2) $X_k \not\subset \bigvee_{i \neq k} X_i$ for every k, $1 \leq k \leq n$.

1.7.7a Theorem *Every Noetherian affine scheme X decomposes into an incontractible quasiunion of a finite number of its primary closed subschemes.*

Proof Let us start with a primary decomposition $X = \bigvee_{1 \leq i \leq n} X_i$, see Remark 1.7.6b. Let Y_j, where $j = 1, \ldots, m$, be the quasiunions of the subschemes X_i with common support. Then $X = \bigvee_{1 \leq i \leq n} Y_j$.

If $Y_1 \subset \bigvee_{2 \leq j \leq n} Y_j$, delete Y_1 from the representation of X. Repeating this process we obtain, after a finite number of steps, a quasiunion $X = \bigvee Y_j$ which satisfies the second condition in the definition of incontractibility. It only remains to verify that the subschemes Y_j are primary.

1.7.7a.i Lemma *The quasiunion of a finite number of primary subschemes with common support is primary and has the same support.*

Proof Let $Y = \bigvee_{1 \leq i \leq n} Y_i$, $\operatorname{supp} Y_i = \operatorname{supp} Y_j$ for all i, j. Let Y_i correspond to an ideal $\mathfrak{a}_i$ in the ring of global functions A on Y. Then $\bigcap_{1 \leq i \leq n} \mathfrak{a}_i = (0)$. Consider a zero divisor $f \in A$. Let $fg = 0$, where $g \neq 0$ and $g \notin \mathfrak{a}_i$ for some i. Since $\mathfrak{a}_i$ is primary, it follows that $f^n \in \mathfrak{a}_i$ for some n. But because $V(\mathfrak{a}_i) = V(0)$, we conclude that $\mathfrak{a}_i$ consists of nilpotents and f is nilpotent; moreover, $\operatorname{supp} Y_i = \operatorname{supp} Y$. $\square$

1.7.8 Theorem (Uniqueness Theorem) *Let $X = \bigvee_{1 \leq i \leq n} X_i$ be an incontractible primary decomposition of a Noetherian affine scheme X. The set of generic points of the irreducible closed sets $\operatorname{supp} X_i$ does not depend on the choice of such a decomposition.*

We denote this set of generic points by $\mathrm{Prime}\,X$ (or $\mathrm{Prime}\,A$ if $X = \mathrm{Spec}\,A$) and call it the *set of prime ideals associated with X* (or A).

We will establish a more precise result giving an invariant characterization of $\mathrm{Prime}\,X$. Let $X = \mathrm{Spec}\,A$ and $X_i = \mathrm{Spec}(A/\mathfrak{a}_i)$ for an ideal $\mathfrak{a}_i \subset A$.

1.7.8a Proposition *The following two statements are equivalent for any reduced scheme X (for general schemes, the statement is only true when $\mathrm{Ann}\,f$ in b) is replaced by $\{g \in A \mid gf \in \bigcap \mathfrak{a}_i\}$):*

(a) *Any prime ideal $\mathfrak{p} \subset A$ corresponds to a generic point of one of the sets $\mathrm{supp}\,X_i$.*
(b) *There exists an element $f \in A$ such that $\mathrm{Ann}\,f := \{g \in A \mid fg = 0\}$ is primary and $\mathfrak{p}$ is its radical.*

Proof (a) $\Longrightarrow$ (b). Let $\mathfrak{p}_j$ be the ideal of a generic point of $\mathrm{supp}\,X_j$. Clearly, $\mathfrak{p}_j = \mathfrak{r}(\mathfrak{a}_j)$, the radical of $\mathfrak{a}_j$, see definition (1.5). Since the representation $X = \bigvee_{1 \le j \le n} X_j$ is incontractible, it follows that $\mathfrak{a}_i \not\supset \bigcap_{j \ne i} \mathfrak{a}_j$, where $1 \le i \le n$. Let us pick an element $f \in (\bigcap_{j \ne i} \mathfrak{a}_j) \setminus \mathfrak{a}_i$ and show that $\mathrm{Ann}\,f$ is primary and $\mathfrak{p}_i$ is its radical.

First of all, $\mathrm{Ann}(f \bmod \mathfrak{a}_i)$ consists only of nilpotents in $A/\mathfrak{a}_i$; therefore, $\mathrm{Ann}\,f \subset \mathfrak{p}_i$ (as $\mathfrak{p}_i$ is the preimage of the nilradical of $A/\mathfrak{a}_j$ under the natural homomorphism $A \longrightarrow A/\mathfrak{a}_i$). Moreover, $\mathfrak{a}_i \subset \mathrm{Ann}\,f$ because, by construction, $f\mathfrak{a}_i \subset \bigcap_j \mathfrak{a}_j = (0)$; therefore, $\mathfrak{r}(\mathrm{Ann}\,f) = \mathfrak{p}_i$.

Now, let us verify that all zero divisors in $A/\mathrm{Ann}\,f$ are nilpotents. Assume the contrary; then there exist elements $g, h \in A$ such that $gh \in \mathrm{Ann}\,f$, $h \notin \mathrm{Ann}\,f$, and g is not nilpotent modulo $\mathrm{Ann}\,f$; hence, g is not nilpotent modulo $\mathfrak{a}_i$, either.

On the other hand, $fgh = 0$; and, because $g \bmod a_i$ is not nilpotent, we see that $fh \bmod a_i = 0$ because a_i is primary, i.e., $fh \in \left(\bigcap_{j \ne i} \mathfrak{a}_j\right) \cap \mathfrak{a}_i = 0$, implying $h \in \mathrm{Ann}\,f$, contrary to the choice of h. This contradiction shows that $\mathrm{Ann}\,f$ is primary.

(b) $\Longrightarrow$ (a). Let $f \in A$ be an element such that $\mathrm{Ann}\,f$ is primary, $\mathfrak{p} = \mathfrak{r}(\mathrm{Ann}\,f)$. Set $\mathfrak{s}_i = (\mathfrak{a}_i : f) := \{g \in A \mid gf \in \mathfrak{a}_i\}$. Since $\bigcap_i \mathfrak{a}_i = (0)$, it is easy to see that $\mathrm{Ann}\,f = \bigcap_i \mathfrak{s}_i$ and $\mathfrak{p} = \mathfrak{r}(\mathrm{Ann}\,f) = \bigcap_i \mathfrak{r}(\mathfrak{s}_i)$. If $f \in \mathfrak{a}_i$, then $\mathfrak{s}_i = \mathfrak{r}(\mathfrak{s}_i) = A$. If, on the contrary, $f \notin \mathfrak{a}_i$, then $\mathrm{Ann}(f \bmod \mathfrak{a}_i)$ consists of nilpotents in $A/\mathfrak{a}_i$, so that $\mathfrak{r}(\mathfrak{s}_i) = \mathfrak{p}_i$.

Therefore, $\mathfrak{p} = \bigcap_{i \in I} \mathfrak{p}_i$, where $I = \{i \mid f \notin \mathfrak{a}_i\}$, implying that $\mathfrak{p}$ coincides with one of the $\mathfrak{p}_i$. Indeed, $V(\mathfrak{p}) = \bigcup_{i \in I} V(\mathfrak{p}_i)$ and $V(\mathfrak{p})$ is irreducible. The proposition is proved, and together with it the uniqueness Theorem 1.7.8. $\square$

1.7.9 Incontractible Primary Decompositions, Continuation 1

An incontractible primary decomposition $X = \bigvee_i X_i$ drastically differs from the decomposition of $\mathrm{supp}\,X$ into the union of its maximal irreducible components: although the supports of the subschemes X_i contain all irreducible com-

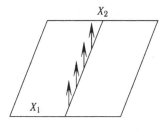

Fig. 1.10

ponents of $\operatorname{supp} X$ only once, the supports may also have another property: $\operatorname{supp} X_i \subset \operatorname{supp} X_j$ for some i, j.

A simple example is given by the ring described in Remark 1.4.8. With the notation of Remark 1.4.8, we have $(0) = (0, \mathfrak{a}) \cap (\mathfrak{p}, 0)$ in A; so that $X = X_1 \vee X_2$, where $\operatorname{supp} X_i = \operatorname{supp} X$ and $\operatorname{supp} X_2 = V((\mathfrak{p}, 0))$. The **space** $\operatorname{supp} X_2$ is entirely contained in the space $\operatorname{supp} X_1$, whereas the sub**scheme** X_2 is distinguished from the "background" by its nilpotents (see Fig. 1.10), where $X_1 = \operatorname{Spec} A$ and $X_2 = \operatorname{Spec} B[T]/(T^2)$, where $B = A/\mathfrak{p}$, and $X = \operatorname{Spec}(\mathbb{R}[T_1, T_2] \oplus \mathfrak{a})$, where $\mathfrak{a} = \mathbb{R}[T_1, T_2]/(T_1)$. Clearly, $\mathfrak{a}^2 = (0)$.

This remark on nilpotents is of general character. Indeed, suppose that in an incontractible decomposition, $\operatorname{supp} X_i \subset \operatorname{supp} X_j$ and $X_i \not\subset X_j$. Then

$$\operatorname{supp}(X_i \cap X_j) = \operatorname{supp} X_i, \quad \text{but } X_i \cap X_j \text{ is a proper subscheme in } X_i.$$

This may only happen when there are more nilpotents in the ring of X_i than in the ring of $X_i \cap X_j$ in which they are induced by nilpotents coming from the bigger space X_j.

Among the components X_i of the incontractible primary decomposition, those for which $\operatorname{supp} X_i$ is maximal are called *isolated*, and the others are called *embedded*. The same terminology is applied to the sets $\operatorname{supp} X_i$ themselves and their generic points, which constitute the set $\operatorname{Prime} X$.

The space of an embedded component may belong simultaneously to several (isolated or embedded) components. Moreover, the chain of components successively embedded into one another may be arbitrarily long.

Therefore, innocent at the first glance, the space of an affine scheme may hide in its depth a complicated structure of embedded primary subschemes, like the one illustrated on Fig. 1.11. The reader should become used to the geometric reality of such a structure.

Of course, when we depict nilpotents by arrows it is impossible to display the fine details of the structure. It is only obvious that on the embedded components the nilpotents grow thicker, thus giving away their presence.

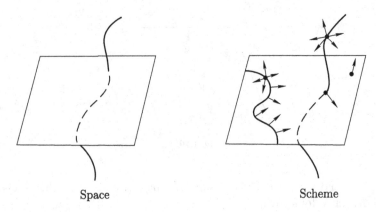

Space   Scheme

Fig. 1.11

1.7.10   Incontractible Primary Decompositions, Continuation 2

The finite set Spec A of prime ideals is invariantly connected with every Noetherian ring A and has a number of important properties. In particular, it enables us to refine Theorem 1.4.8

1.7.10a Theorem *A given element $f \in A$ is a zero divisor if and only if it vanishes (as a function) on one of the components of the incontractible primary decomposition of* Spec A.

In other words, the set of zero divisors of A coincides with $\bigcap_{\mathfrak{p} \in \mathrm{Prime}\,A} \mathfrak{p}$.

Proof First, let us show that if $f \notin \bigcap_{\mathfrak{p} \in \mathrm{Prime}\,A} \mathfrak{p}$, then $\mathrm{Ann} f = (0)$.

Let $(0) = \bigcup_i \mathfrak{a}_i$ be the incontractible primary decomposition, where $\mathfrak{p}_i = \mathfrak{r}(\mathfrak{a}_i)$. Let $fg = 0$. Since $f \notin \mathfrak{p}_i$, it follows that $\mathfrak{a}_i$ being primary implies that $g \in \mathfrak{a}_i$. This is true for all i; therefore, $g = 0$.

Conversely, let $\mathrm{Ann}(f) = (0)$. Suppose that $f \in \mathfrak{p}_i$; because A is Noetherian, it follows that $f^k \in \mathfrak{p}_i^k \subset \mathfrak{a}_i$ for some $k \geq 1$; i.e., $(\mathfrak{a}_i : (f^k)) = A$.[17] On the other hand, $\mathrm{Ann}(f^k) = (0)$, and so

$$(0) = \mathrm{Ann}(f^k) = \bigcap_j (\mathfrak{a}_j : (f^k)) = \bigcap_{j \neq i} (\mathfrak{a}_j : (f^k)) \supset \bigcap_{j \neq i} \mathfrak{a}_j.$$

This contradicts the incontractibility of the decomposition; end of the proof.   □

---

[17]Recall that the *quotient* $(\mathfrak{a} : \mathfrak{b})$ *of two ideals* in A is defined to be the set $\{x \in A \mid x\mathfrak{b} \subset \mathfrak{a}\}$. It is easy to see that $(\mathfrak{a} : \mathfrak{b})$ is an ideal in A. In particular, the ideal $(0 : \mathfrak{b})$ is called the *annihilator* of $\mathfrak{b}$ and is denoted $\mathrm{Ann}(\mathfrak{b})$.

1.7.11   Incontractible Primary Decompositions for Noetherian Schemes

Finally, notice that the uniqueness Theorem 1.7.8 only concerns the supports of primary components of incontractible decompositions, not the components themselves. About the latter one can only claim the following.

1.7.11a Theorem *The set of isolated components of an incontractible primary decomposition of a Noetherian scheme* Spec *A does not depend on the choice of the decomposition.*

For embedded components this statement is false.
We skip the proof; the reader may find it in [ZS, v. 1, Theorem 4.5.8] or [Lang].

1.7.12 Exercises

(1) Let K be an algebraically closed field.

   (a) Describe all primary closed subschemes of the line Spec $K[T]$.
   (b) Same for non-closed fields.
   (c) Same for Spec $\mathbb{Z}$.

(2) Describe, up to an isomorphism, the primary closed subschemes of the plane Spec $K[T_1, T_2]$ with support at $V(T_1, T_2)$ and such that their local rings are of length ≤ 3.

1.8   Hilbert's Nullstellensatz (Theorem on Zeroes)

In this section we establish that the closed subschemes of finite-dimensional affine spaces over any field or over $\mathbb{Z}$ have many closed points.

1.8.1 Theorem *Let A be any ring of finite type. Then* Spm *A is dense in* Spec *A.*

1.8.1a Corollary *The intersection $X \cap$ Spm A is dense in X for any open or closed subset $X \subset$ Spec A.*

Indeed, if $X = V(E)$ and we identify X with Spec $A/(E)$, then Spm $A \cap V(E)$ coincides with Spm$(A/(E))$ and $A/(E)$ is a ring of finite type together with A. This easily implies the statement for open sets, too.   □

The space Spm A is easier to visualize because it has no nonclosed points (the "big open sets" still remain nevertheless). On the other hand, Corollary 1.8.1a implies that for rings of finite type the space of Spec A is uniquely recovered from Spm A (assuming that the induced topology in Spm A is known).

The recipe is as follows:

(1) The points of Spec A are in a one-to-one correspondence with the irreducible closed subsets of Spm A. (Hence, to every irreducible closed subset of Spm A there corresponds its "generic point" in Spec A.)

(2) Every closed subset of Spec A consists of generic points of all irreducible closed subsets of a closed subset of Spm A.

We advise the reader to prove these statements in order to understand them :-)

Proof of Theorem 1.8.1 We will successively widen the class of rings for which this theorem is true.

(a) *Let K be algebraically closed.*

Then the set of closed points of Spec $K[T_1, \ldots, T_n]$ *is dense.*

The closure of the set of closed points coincides with the space of zeroes of all functions which vanish at all closed points, and therefore it suffices to prove that a polynomial F belonging to all maximal ideals of $K[T_1, \ldots, T_n]$ is identically zero. But for such a polynomial we have $F(t_1, \ldots, t_n) = 0$ for all $t_1, \ldots, t_n \in K$, and an easy induction on n shows that $F = 0$ (here we actually use not even closedness, but only the fact that K is infinite).

(b) *The same claim as in* (a), *but K is not supposed to be algebraically closed.*

Denote by $\overline{K}$ the algebraic closure of K. We have a natural morphism

$$i: A = K[T_1, \ldots, T_n] \hookrightarrow \overline{K}[T_1, \ldots, T_n] = B.$$

The ring B is integral over K, and therefore thanks to Theorems 1.6.5 and 1.6.6, we have

$$\overline{\mathrm{Spm}\, A} = {}^a i(\overline{\mathrm{Spm}\, B}) = {}^a i(\mathrm{Spec}\, B) = \mathrm{Spec}\, A.$$

(c) *Theorem 1.8.1 holds for the rings A of finite type without zero divisors over K.*

Indeed, by Noether's normalization theorem, see Theorem 1.5.3f, there exists a polynomial subalgebra $B \subset A$ such that A is integral over B. By the already proved, $\overline{\mathrm{Spm}\, B} = \mathrm{Spec}\, B$, and literally the same argument as in (b) shows that $\overline{\mathrm{Spm}\, A} = \mathrm{Spec}\, A$.

(d) *Theorem 1.8.1 holds for any rings A of finite type over a field.*

Indeed, any irreducible component of Spec A is homomorphic to Spec $A/\mathfrak{p}$, where $\mathfrak{p}$ is a prime ideal. The ring $A/\mathfrak{p}$ satisfies the conditions of (c), hence, the closed points are dense in all irreducible components of Spec A, and therefore in the whole space.

(e) *The same as in case* (d) *for the rings A of finite type over $\mathbb{Z}$.*

1.8.2 Lemma *No field of characteristic 0 can be a finite type algebra over $\mathbb{Z}$.*

Proof of Lemma 1.8.2 If the field K of characteristic 0 is a finite type algebra over $\mathbb{Z}$, then thanks to Noether's normalization theorem 1.5.3f, there exist algebraically independent over $\mathbb{Q}$ elements $t_1, \ldots, t_r \in K$ such that K is integral over the algebra $R = \mathbb{Q}[t_1, \ldots, t_r]$.

By Proposition 1.6.6, the natural map $\operatorname{Spm} K \longrightarrow \operatorname{Spm} \mathbb{Q}[t_1, \ldots, t_r]$ is surjective; further, since $\operatorname{Spec} K$ contains only one closed point, so does $\operatorname{Spec} R$, which is only possible if $R = \mathbb{Q}$. Therefore, K is integral over $\mathbb{Q}$, and hence K is a finite extension of the field $\mathbb{Q}$.

Let $x_1, \ldots, x_n$ generate K as a $\mathbb{Z}$-algebra. Each of the x_j is a root of a polynomial with rational coefficients. If N is the LCM of the denominators of these coefficients, then, as is easy to see, all Nx_j are integral over $\mathbb{Z}$; if y is the product of m of the generators $x_1, \ldots, x_n$ (possibly, with multiplicities), then $N^m y$ is integral over $\mathbb{Z}$. Since all elements of K are linear combinations of such products with integer coefficients, it follows that for any $y \in K$, there exists a natural m such that $N^m y$ is integral over $\mathbb{Z}$.

Now let p be a prime number that does not divide N. Since $1/p \in \mathbb{Q} \subset K$, we see that the non-integer rational number N^m/p is integral over $\mathbb{Z}$; but Exercise 1.6.7 shows that this is impossible. □

To complete the proof of Theorem 1.8.1 in case (e), we denote by $\varphi: \mathbb{Z} \longrightarrow A$ the natural homomorphism and show that $^a\varphi(\operatorname{Spm} A) \subset \operatorname{Spm} \mathbb{Z}$. Indeed, otherwise there exists a maximal ideal $\mathfrak{p} \subset A$ such that $\varphi^{-1}(\mathfrak{p}) = (0)$; so $\mathbb{Z}$ can be embedded into the field $A/\mathfrak{p}$ (and hence this field is of characteristic 0), which leads to a contradiction with Lemma 1.8.2.

Thus, $\operatorname{Spm} A = \bigcup_p V(p)$, where p runs over primes. Since $V(p)$ is homeomorphic to $\operatorname{Spec} \mathbb{Z}/p\mathbb{Z}$, which is an algebra of finite type over A/pA, we see that the set of closed points of $V(p)$ is dense in $\operatorname{Spec} A/pA$; hence, it is also dense in $\operatorname{Spec} A$. □

1.8.3 Proposition *Let $\mathfrak{p} \subset A$ be a maximal ideal in a ring of finite type over $\mathbb{Z}$ (resp. a field K). Then $A/\mathfrak{p}$ is a finite field (resp. a finite algebraic extension of K).*

Proof It follows from step (e) of the proof of Theorem 1.8.1 that it suffices to consider the case of a ring A over a field K. The quotient ring $A/\mathfrak{p}$ modulo the maximal ideal, being a field, contains a unique maximal ideal. On the other hand, by Noether's normalization theorem, see Theorem 1.5.3f, A is an integral extension of the polynomial ring B in n indeterminates over K. The case $n \geq 1$ is impossible, because then B, and therefore A, would have had infinitely many maximal ideals. Therefore, $n = 0$, and A is an integer extension of finite type of K. This proves the statement. □

1.8.4 Hilbert's Nullstellensatz

Consider the case where K is algebraically closed. By Proposition 1.8.3, the closed points of $\operatorname{Spec} A$ are, in this case, in one-to-one correspondence with the K-points of the scheme $\operatorname{Spec} A$; the space of the latter points is called an *affine algebraic variety* over K in the classical sense of the word. The discussion of Sect. 1.8.1 shows that in this case $\operatorname{Spec} A$ with the spectral topology and the space of geometric

K-points of the scheme Spec A with the Zariski topology are essentially equivalent notions: the passage from one to the other requires no additional data.

Finally, let us give the classical formulation of *Hilbert's Nullstellensatz* in the language of systems of equations.

1.8.4a Theorem (Hilbert's Nullstellensatz) *Let K be algebraically closed, $F_i \in K[T]$ for $i \in I$ be polynomials in $T = (T_1, \ldots, T_n)$, and $F = (F_i)_{i \in I}$.*

(a) *The system of equations $F_i = 0$ for $i \in I$, has a solution in K if and only if the equation $1 = \sum_i F_i X_i$ has no solutions in $K[T_1, \ldots, T_n]$, i.e., the ideal (F) does not coincide with the whole ring.*

(b) *If a polynomial $G \in K[T_1, \ldots, T_n]$ vanishes on all the solutions of the system of equations $F_i = 0$ for $i \in I$, then $G^n = \sum_i F_i G_i$, where $G_i \in K[T_1, \ldots, T_n]$ for some positive integer n.*

Proof
(a) If (F) does not coincide with the whole ring, then, by Theorem 1.2.3, we have Spec $K[T_1, \ldots, T_n]/(F) \neq \emptyset$. Therefore, the spectrum contains a maximal ideal for which the corresponding residue field coincides, thanks to Proposition 1.8.3, with K, and the images of T_i in this field give a solution of the system of equations $F_i = 0$ for $i \in I$.

(b) If G vanishes at all solutions of the system of equations, then the image of G in $K[T_1, \ldots, T_n]/(F)$ belongs to the intersection of all maximal ideals of this ring, hence, by Theorem 1.7.10, to the intersection of all prime ideals. Therefore, G is nilpotent thanks to Theorem 1.4.7. $\square$

1.9   Fiber Products

This section contains no significant theorems. We only give a construction of the fiber product of affine schemes. This notion, though simple, is among the most fundamental ones and explains in part the popularity of the tensor product in modern commutative algebra. Our main aim is to connect the fiber products with intuitive geometrical images.

We advise the reader to refresh the knowledge of categories (see Sect. 1.16) before reading this section.

1.9.1   Fiber Product

Let C be a category, $S \in \text{Ob}\,C$, and C_S the category of "objects over S". The *fiber product* of two objects $\psi: Y \longrightarrow S$ and $\varphi: X \longrightarrow S$ over S is their product in C_S.

In other words, the *fiber product* of $\psi: Y \longrightarrow S$ and $\varphi: X \longrightarrow S$ is a triple (Z, π_1, π_2), where $Z \in \mathrm{Ob}\, \mathsf{C}$, $\pi_1: Z \longrightarrow X$, $\pi_2: Z \longrightarrow Y$, such that

(a) the diagram (1.10), where $(Z, \varphi\pi_1) = (Z, \psi\pi_2) \in \mathrm{Ob}\, \mathsf{C}_S$ and $\pi_1, \pi_2 \in \mathrm{Mor}\, \mathsf{C}_S$, commutes;
(b) for any object $X: Z' \longrightarrow S$ (in what follows it will be denoted simply by Z') of C_S, the maps induced by the morphisms π_1, π_2 identify the set $\mathrm{Hom}_{\mathsf{C}_S}(Z', Z)$ with $\mathrm{Hom}_{\mathsf{C}_S}(Z', X) \times \mathrm{Hom}_{\mathsf{C}_S}(Z', Y)$.

In still other words, (Z, π_1, π_2) is a universal object in the class of all triples that make the diagram (1.10) commute.

$$
\begin{array}{ccc}
Z & \overset{\pi_1}{\longrightarrow} & X \\
{\scriptstyle \pi_2}\downarrow & & \downarrow{\scriptstyle \varphi} \\
Y & \underset{\psi}{\longrightarrow} & S
\end{array}
\tag{1.10}
$$

The diagram (1.10) with the properties (a) and (b) is sometimes called a *Cartesian square*. The object Z in it is usually denoted by $X \times_S Y$ and is called the *fiber product* of X and Y over S. When using this short notation one should not forget that the four morphisms $X \longrightarrow S$, $Y \longrightarrow S$, $X \times_S Y \longrightarrow X$, and $X \times_S Y \longrightarrow Y$, the first two of which are vital, are not explicitly indicated in it.

Formally speaking, the usual—set-theoretical—direct product is not a particular case of the fiber product; but it is so if the category C has a final object F. Then $X \times_F Y$ is essentially the same as $X \times Y$.

The fiber product exists in Sets; we will elucidate its meaning with several examples. The next lemma is evident.

1.9.1i Lemma *Let $\varphi : X \longrightarrow S$ and $\psi: Y \longrightarrow S$ be maps of sets; define*

$$
Z := \{(x, y) \in X \times Y \mid \varphi(x) = \psi(y)\} \subset X \times Y
$$

and define $\pi_1: Z \longrightarrow X$ and $\pi_2: Z \longrightarrow Y$ as the maps induced by the projections $X \times Y \longrightarrow X$ and $X \times Y \longrightarrow Y$, respectively. Then the triple (Z, π_1, π_2) is a fiber product of X and Y over S.

This construction explains where the name of our operation stems from: over every point of S the fiber of the map $Z \longrightarrow S$ is the direct product of the fibers of X and Y.

1.9.2 Example The following notions are ubiquitous.

Product In Sets, the one-point set $*$ is a final object, and so

$$
X \times_* Y = X \times Y \quad \text{for any } X, Y \in \mathsf{Sets}.
$$

Intersection Let φ and ψ be embeddings of X and Y, as subsets, into S. Then identifying $Z = X \times_S Y$ with a subset of S we see that $Z = X \cap Y$.

The Fiber of a Map Let $Y = F$ be a final object, $\psi(F) = s \in S$. Then $Z = \varphi^{-1}(s)$. More generally, if ψ is an embedding, then upon identifying Y with $\psi(Y) \subset S$ we obtain $Z = \varphi^{-1}(Y)$.

Change of Base This terminology comes from topology: if $\varphi: X \longrightarrow S$ is a bundle (in any sense) and $\psi: S' \longrightarrow S$ a morphism of topological spaces, then the bundle $X' = X \times_S S'$ is said to be obtained from $X \longrightarrow S$ by *changing the base S to S'*. The other name for it is the *induced bundle*.

1.9.3   Fiber Products Exist

The above examples will serve us as a model for the corresponding notions in the category of schemes (affine ones, for the time being). First of all, let us establish their existence.

1.9.3a Theorem *Let $X = \operatorname{Spec} A$, $Y = \operatorname{Spec} B$, $S = \operatorname{Spec} C$, where A and B are C-algebras. The fiber product of X and Y over S exists and is represented by the triple $(\operatorname{Spec} A \otimes_C B, \pi_1, \pi_2)$, where π_1 (resp. π_2) is the map induced by the C-algebra homomorphism $A \longrightarrow B \otimes_C B$ such that $f \longmapsto f \otimes 1$ (resp. $B \longrightarrow A \otimes_C B$ such that $g \longmapsto 1 \otimes g$).*

For a proof, see [Lang], where the fact that fiber coproducts in the category **Rings** exist, and are described just as stated, is established. The inversion of arrows gives the statement desired.   □

Observe that the category of affine schemes has a final object, $\operatorname{Spec} \mathbb{Z}$. So we may speak about the *absolute product*

$$X \times Y := \operatorname{Spec} A \otimes_{\mathbb{Z}} B.$$

1.9.4   Warning

The statement "*the set of points $|X \times_S Y|$ of the scheme $X \times_S Y$ is the fiber product $|X| \times_S |Y|$ of the sets of points $|X|$ and $|Y|$ over $|S|$*" (here $|X|$ denotes the set of points of X, not the cardinality of this set) is only true for the points with values in C-algebras; i.e., when $S = \operatorname{Spec} C$ or, which is the same, for the sets of morphisms over S. Here are typical examples showing what can happen otherwise.

1.9.4a Examples
(1) Let K be a field, $S = \operatorname{Spec} K$; let $X = \operatorname{Spec} K[T_1]$ and $Y = \operatorname{Spec} K[T_2]$, with the obvious morphisms between these spectra. Then $X \times_S Y$ is the plane over K; it has an abundance of non-closed points: generic points of irreducible curves non-parallel to the axes, which are not representable by pairs (x, y), with $x \in X$, $y \in Y$.

(2) Let $L \supset K$ be fields.

(2a) Let $L \rhd K$ be a finite Galois field extension, $X = \operatorname{Spec} L$, and $S = \operatorname{Spec} K$. Let us describe $X \times_S X$ or, dually, $L \otimes_K L$.

Let us represent the second factor L as $L = K[T]/(F(T))$, where $F(T)$ is an irreducible polynomial. In other words, in L, we take a primitive element $t = T \bmod (F)$ over K.

It follows from the definition of the tensor product that in this case, as L-algebras,

$$L \otimes_K L \simeq L[T]/(F(T))$$

if we assume that the L-algebra structure on $L \otimes_K L$ is determined by the map $l \longmapsto l \otimes 1$. But, by assumption, $F(T)$ factorizes in $L[T]$ into linear factors: $F(T) = \prod_{1 \le i \le n}(T - t_i)$, where the t_i are all the elements conjugate to t over K and $n = [L : K]$.

By the general theorem on the structure of modules over principal ideal rings, see [Lang], we obtain

$$L \otimes_K L \simeq L[T]/\left(\prod(T - t_i)\right) \simeq \prod L[T]/(T - t_i) \simeq L^n.$$

In particular, $\operatorname{Spec} L \otimes_K L \simeq \coprod_{1 \le i \le n} \operatorname{Spec} L$: though $\operatorname{Spec} L$ consists of only one point, there are, miraculously, n points in $\operatorname{Spec} L \otimes_K L$.

(2b) Trouble of another nature may arise if for L we take a purely nonseparable extension of K. Let, e.g., $F(T) = T^p - g$, where $g \in K \setminus K^p$ for $K^p := \{x^p \mid x \in K\}$ and $p = \operatorname{Char} K$. Then in $L[T]$ we have $T^p - g = (T - t)^p$, where $t = g^{1/p}$, so that

$$L \otimes_K L \simeq L[T]/((T - t)^p) \simeq L[T]/(T^p);$$

therefore we have acquired nilpotents which were previously lacking. The space of $\operatorname{Spec} L \otimes_K L$ consists, however, of a single point. □

1.9.5 Examples

Let us give examples which are completely parallel to the set-theoretical construction.

(1) Let $X = \operatorname{Spec} A$, let $Y_1 \xrightarrow{i_1} X$ and $Y_2 \xrightarrow{i_2} X$ be two *closed* subschemes of X determined by ideals $\mathfrak{a}_1, \mathfrak{a}_2 \subset A$. Thanks to results of Sect. 1.7.2, their intersection $Y_1 \cap Y_2$ represents the functor $Y_1(Z) \cap Y_2(Z)$, i.e., should coincide with their fiber product over X. This is indeed the case: the corresponding

statement on rings is

$$A/(\mathfrak{a}_1 + \mathfrak{a}_2) \simeq A/\mathfrak{a}_1 \otimes_A A/\mathfrak{a}_2$$

and is easy to verify directly.

(2) Let $\operatorname{Spec} B = Y \longrightarrow X = \operatorname{Spec} A$ be a morphism of affine schemes, $\overline{k(x)}$ the algebraic closure of the field of fractions corresponding to a given point $x \in X$. The natural homomorphism $A \longrightarrow \overline{k(x)}$ represents a geometric point with center at x. The fiber product $Y_x = Y \times_X \operatorname{Spec} \overline{k(x)}$ is called the *geometric fiber* of Y over x, and $Y \times_X \operatorname{Spec} k(x)$ is called the *ordinary fiber*.

(2a) A particular case: $\operatorname{Spec} A/pA$ is the fiber of $\operatorname{Spec} A$ over $(p) \in \operatorname{Spec} \mathbb{Z}$ for any prime p and any ring A.

1.9.6   The Diagonal

Let $\operatorname{Spec} B = X \longrightarrow S = \operatorname{Spec} A$ be a morphism of affine schemes. The commutative diagram

$$
\begin{array}{ccc}
X & \xrightarrow{\ \text{id}\ } & X \\
{\scriptstyle \text{id}}\downarrow & & \downarrow \\
X & \longrightarrow & S
\end{array}
$$

defines a morphism $\delta: X \longrightarrow X \times_S X$ (see Sect. 1.9.1b), called the *diagonal* morphism.

1.9.6a Proposition *The diagonal morphism δ identifies X with the closed subscheme Δ_X of $X \times_S X$ determined by the ideal*

$$I_{\Delta_X} = \operatorname{Ker}(\mu: B \otimes_A B \longrightarrow B), \text{ where } \mu(b_1 \otimes b_2) = b_1 b_2.$$

Proof Writing down all necessary diagrams we see that $\delta = {}^a\mu$. Since μ is surjective, we see that its kernel determines a closed subscheme isomorphic to the image of δ.   □

The scheme Δ_X is said to be the (*relative* over S) *diagonal*.

1.10 Vector Bundles and Projective Modules

1.10.1 Families of Vector Spaces

Let $\psi: Y \longrightarrow X$ be a morphism of affine schemes, $X = \operatorname{Spec} A$, $Y = \operatorname{Spec} B$, and $\varphi: A \longrightarrow B$ the corresponding ring homomorphism. We would like to single out a class of morphisms which is similar to the class of *locally trivial vector bundles* in topology.

It is convenient to start with the wider notion of "family of vector spaces" (a term borrowed from [A]). Example 1.2.1 shows that an analogue of a vector space V over a field K is given by the scheme $\operatorname{Spec} S_K^{\bullet}(V^*)$, where $V^* = \operatorname{Hom}_K(V, K)$ and $S_K^{\bullet}(V^*)$ is the symmetric algebra of V^*. Replacing here K by an arbitrary ring A and V by an A-module M, we arrive at the following definition.

With the above notation, let $\chi: M \longrightarrow B$ be an A-module morphism. Suppose χ induces an A-algebra isomorphism $S_A^{\bullet}(M) \xrightarrow{\simeq} B$. Then the pair (χ, ψ) is said to be a *family of vector spaces* over $X = \operatorname{Spec} A$ and M is said to be the *module that defines the family*.

The *morphisms of families of vector spaces* over a fixed base are naturally defined. The category of such families is dual to the category of A-modules; so, in particular, every family of vector spaces is determined by its module M up to an isomorphism.

The following notion is more important for us at the moment.

1.10.2 Families of Vector Spaces Exist

With the above notation, let there be given a ring homomorphism $A \longrightarrow A'$ that defines a scheme morphism

$$X' = \operatorname{Spec} A' \longrightarrow X = \operatorname{Spec} A.$$

Consider a family of vector spaces (χ', ψ'), where

$$\chi' = \operatorname{id} \otimes \chi: M' = A' \otimes_A M \longrightarrow A' \otimes_A B$$

and ψ' is a morphism $Y' = \operatorname{Spec} A' \otimes_A B \longrightarrow X'$. This family is said to be *induced by changing the base X' to X*.

We see that X' is indeed a family of vector spaces, because there is a canonical isomorphism

$$S_{A'}^{\bullet}(A' \otimes_A M) \xrightarrow{\simeq} A' \otimes_A S_A^{\bullet}(M). \tag{1.11}$$

In particular, if A' is a field, then Y' is the scheme of the vector space $(A' \otimes_A M)^*$ over A'. This means that all fibers of the family $\psi: Y \longrightarrow X$ over geometric points are vector spaces, which justifies the name "scheme of the vector space". The dimensions of the fibers can, of course, undergo jumps.

Note also that thanks to the isomorphism (1.11) the scheme Y' is identified with the fiber product $X' \times_X Y$, so that our change of base operation is an exact analogue of the topological one.

1.10.3   Vector Bundles

A family of vector spaces $\chi: Y \longrightarrow X$ is said to be *trivial* if its defining A-module is free.

The families of vector spaces trivial in a neighborhood of any point are naturally referred to as *vector bundles*. It is not quite clear though how to define local triviality: the neighborhoods of points in $\operatorname{Spec} A$ are just topological spaces, not schemes. Here we first encounter a problem that will be systematically investigated in the next chapter. For our intermediate objectives, it is natural to adopt the following quick fix.

A family of vector spaces $\chi: Y \longrightarrow X$ is said to be *trivial at* $x \in X$ if there exists an open neighborhood $U \ni x$ such that, for any morphism $\psi: X' \longrightarrow X$ with $\psi(X') \subset U$, the induced family $Y' \longrightarrow X'$ is trivial.

We will now replace this triviality condition with another condition easier to verify. First of all, because the big open sets $D(f)$ constitute a basis of the topology of $\operatorname{Spec} A$, it suffices to consider neighborhoods of the form $D(f)$. They possess the following remarkable property.

1.10.3a Proposition *Let A be a ring, $f \in A$ not a nilpotent. Set $X_f := \operatorname{Spec} A_f$ and denote by $i^*: X_f \longrightarrow X$ the morphism induced by the homomorphism $i: A \longrightarrow A_f$, $g \longmapsto g/1$. Then*

(a) *i^* determines a homeomorphism $X_f \simeq D(f)$*
(b) *For any morphism $\psi: X' \longrightarrow X$ such that $\psi(X') \subset D(f)$, there exists a unique morphism $\chi: X' \longrightarrow X_f$ such that the diagram*

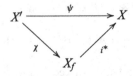

commutes.

This implies that the family of vector spaces $Y \longrightarrow X$ is trivial at $x \in \operatorname{Spec} A = X$ if and only if, for any element $f \in A$ such that $f(x) \neq 0$, the family induced over X_f is trivial. Translating this into the language of modules we find a simple condition which will be used in what follows.

1.10.3a.i Corollary *An A-module M determines a family of vector spaces trivial at $x \in \operatorname{Spec} A$ if and only if there exists $f \in A$ such that $f(x) \neq 0$ and such that the A_f-module $M_f := A_f \otimes_A M$ is free.*

Any module M satisfying all the conditions of Corollary 1.10.3a.i for all points $x \in \operatorname{Spec} A$ is said to be *locally free*.

Proof of Proposition 1.10.3a

(a) This is a particular case of Theorem 1.6.4f.
(b) This statement expresses the known universality property of the rings of fractions. Indeed, let $\psi : A \longrightarrow A'$ be a ring homomorphism such that $^a\psi(\operatorname{Spec} A') \subset D(f)$. This means that f does not belong to any of the ideals $\psi^{-1}(\mathfrak{p})$, where $\mathfrak{p} \in \operatorname{Spec} A'$, i.e., $\psi(f)$ does not vanish on $\operatorname{Spec} A'$. Therefore, $\psi(f)$ is invertible in A'.

In the category of such A-algebras, the morphism $A \longrightarrow A_f$ is a universal object (see Theorem 1.6.4c), which proves the statement desired.

In particular, if $D(f) = D(g)$, then the ring A_f is canonically isomorphic to A_g.

$\square$

1.10.4 Main Definition and Main Result of This Section

A *vector bundle over a scheme* $X = \operatorname{Spec} A$ is a family of vector spaces which is locally trivial at each point of $\operatorname{Spec} A$.

Unless otherwise stated, until the end of this section we will consider only Noetherian rings and modules.

Recall that a module M is said to be *projective* if it is isomorphic to a direct summand of a free module.

1.10.4a Theorem *An A-module M determines a vector bundle over $\operatorname{Spec} A$ if and only if it is projective.*

The theorem claims that the class of locally free modules coincides with the class of projective modules. This is precisely the statement we will prove; first the inclusion in one direction and then in the opposite one. We have to perform a rather hard job and we use the opportunity to establish en route more auxiliary results than is actually needed at this point: they will serve us later.

1.10.5 Localizations of Modules

Let $S \subset A$ be a multiplicative set that does not contain 0, and M an A-module.

Define the localization M_S, sometimes denoted by $M[S^{-1}]$, to be the set $\{(m, s) \mid m \in M,\ s \in S\}/R$, where the relation R is given by the rule

$$(m, s) \sim (m', s') \iff t(s'm - sm') = 0 \text{ for some } t \in S.$$

In particular, for $M = A$, we get the definition of A_S.

The A-module M_S is naturally endowed with an A_S-module structure:

$$\frac{m_1}{s_1} + \frac{m_2}{s_2} = \frac{s_2 m_1 + s_1 m_2}{s_1 s_2} \quad \text{and} \quad \frac{a}{s}\frac{m}{t} = \frac{am}{st}$$

$$\text{for any } s_i \in S,\ a \in A,\ m_i, m \in M.$$

Clearly, $M_S = A_S \otimes_A M$.

1.10.5a Lemma *The equality $m/s = 0$ holds if and only if there exists $t \in S$ such that $tm = 0$. In particular, the kernel of the natural homomorphism*

$$M \longrightarrow M_S, \qquad m \longmapsto m/1,$$

consists of the elements m such that $(\mathrm{Ann}\, m) \cap S \neq \emptyset$.

Proof Clearly, $tm = 0 \implies tm/ts = 0 = m/s$. To prove the converse implication, consider first a particular case.

(a) *M is free.* Let $\{m_i\}_{i \in I}$ be a free A-basis of M. Then $\{m_i = m_i/1\}_{i \in I}$ is a free A_S-basis of M_S. Let $m = \sum f_i m_i$, where $f_i \in A$, be an element of M. If $m/s = 0$, then $f_i/s = 0$ for all i, and hence there exist $t_i \in S$ such that $t_i f_i = 0$. Let $t := \prod t_i$, where i runs over the finite subset of indices $I_0 \subset I$ for which $f_i \neq 0$. Clearly, $tm = 0$ because $t f_i = 0$ for all $i \in I_0$.

(b) *The general case.* There exists an exact sequence

$$F_1 \xrightarrow{\varphi} F_0 \xrightarrow{\psi} M \longrightarrow 0,$$

where F_0, F_1 are free modules. Tensoring the sequence by A_S we get the exact (see [Lang]) sequence

$$(F_1)_S \xrightarrow{\varphi_S} (F_0)_S \xrightarrow{\psi_S} M_S \longrightarrow 0,$$

where $\varphi_S = \mathrm{id}_{A_S} \otimes_A \varphi$, and so on.

Let $m/s = 0$ for $m = \psi(n)$, where $n \in F_0$. Then $\psi_S(n/s) = 0$; this implies

$$n/s = \varphi_S(l/t) = \varphi(l)/t, \text{ where } l \in F_1 \text{ and } t \in S.$$

In other words, $(tn - s\varphi(l))/st = 0$ in $(F_0)_S$. Since F_0 is free, it follows that there exists $r \in S$ such that $rtn = rs\varphi(l)$ in F_0. Applying ψ to this relation we get

$$rtm = \psi(rtn) = rs\psi \circ \varphi(l) = 0,$$

as desired. $\square$

Observe that we did not use the Noetherian property.

1.10.5b Corollary *Let M be a Noetherian A-module, $f \in A$. There exists an integer $q > 0$ such that $f^q m = 0$ for all $m \in \mathrm{Ker}(M \longrightarrow M_f)$.*

Proof Select the needed value q_i for each member of a finite family of generators of the kernel, and set $q = \max\limits_i q_i$. $\square$

1.10.6   Tensoring Exact Sequences

In the proof of Lemma 1.10.5 we have used the following general property of the tensor product:

> *Tensoring sends short exact sequences into sequences exact everywhere except at the leftmost term.*

Tensoring by A_S, however, possesses a stronger property: it completely preserves exactness; this means A_S is what is called a *flat A-algebra*.

1.10.6a Proposition *The sequence $M_S \xrightarrow{\varphi_S} N_S \xrightarrow{\psi_S} P_S$ of A_S-modules is exact for any exact sequence $M \xrightarrow{\varphi} N \xrightarrow{\psi} P$ of A-modules.*

Proof $\psi \circ \varphi = 0 \Longrightarrow \psi_S \circ \varphi_S = 0 \Longrightarrow \mathrm{Ker}\,\psi_S \supset \mathrm{Im}\,\varphi_S$.

Conversely, let $n/s \in \mathrm{Ker}\,\psi_S$; then $\psi(n/s) = 0$ implying, thanks to the above, $t\psi(n) = 0$ for some $t \in S$. Therefore, $tn = \varphi(m)$, and so

$$n/s = tn/ts = \varphi(m)/ts = \varphi_S(m/ts),$$

as desired. $\square$

1.10.7   Lifts of A_f-module Homomorphisms

Let $\varphi: M \longrightarrow N$ be an A-module homomorphism. For any $f \in A$, we have an induced A_f-module homomorphism $\varphi_f = \mathrm{id}_{A_f} \otimes \varphi: M_f \longrightarrow N_f$. In this situation we will say the morphism $\psi: M_f \longrightarrow N_f$ can be lifted to $\varphi: M \longrightarrow N$ if $\varphi_f = \psi$.

1.10.7a Lemma *Let F be a free Noetherian A-module, M a Noetherian A-module, and $f \in A$ such that M_f a free A-module.*

Then, for any homomorphism A_f-module homomorphism $\varphi: M_f \longrightarrow F_f$, there exists an integer q such that the homomorphism $f^q \varphi: M_f \longrightarrow F_f$ can be lifted to a homomorphism $M \longrightarrow F$.

Proof First of all, F_f is free and has a finite number of generators, so that φ is given by a finite number of coordinate A_f-morphisms $M_f \longrightarrow A_f$. If being multiplied by an appropriate power of f each of them can be lifted to a morphism $M \longrightarrow A$, then so does φ. Therefore, we may and will assume that $F = A$.

Let m_i, where $i = 1, \ldots, n$, be a family of generators of M. Multiplying φ by an appropriate power of f, we may assume that $\varphi(m_i) = g_i/1$, where $g_i \in A$ for all i.

It is tempting to lift $\varphi: M_f \longrightarrow A_f$ to a homomorphism $\psi: M \longrightarrow A$ by setting $\psi(m_i) = g_i$. This, however, might prove impossible because there are relations $\sum f_i m_i = 0$ such that $\sum f_i g_i \neq 0$. But we have the equality $\sum f_i(g_i/1) = 0$, and therefore, thanks to Lemma 1.10.5, the set

$$\left\{ \sum f_i g_i \mid \sum f_i m_i = 0 \right\}$$

constitutes a Noetherian A-submodule of $\mathrm{Ker}(A \longrightarrow A_f)$. By Corollary 1.10.5, this submodule is annihilated by f^q for some q. This implies that there exists a homomorphism $f^q \psi: M \longrightarrow A$ such that $f^q \psi(m_i) = f^q g_i$ because $(f^q \psi)_f = f^q \varphi$.
$\qquad\square$

1.10.8   Locally Free Modules Are Projective

Now, we can establish the "only if" part of Theorem 1.10.4.

1.10.8a Proposition *Locally free modules are projective.*

Proof Let M be a Noetherian locally free A-module, $\psi: F \longrightarrow M$ an epimorphism, where F is a Noetherian free module. In order to prove that M is a direct summand of F, we have to find a section, i.e., an A-module homomorphism $\varphi: M \longrightarrow F$ such that $\psi \circ \varphi = \mathrm{id}_M$. More generally, let

$$P = \{\chi \in \mathrm{Hom}_A(M, M) \mid \chi = \psi \circ \varphi \text{ for some } \varphi \in \mathrm{Hom}_A(M, F)\}.$$

First, let us show that, for every point $x \in \operatorname{Spec} A$, there exists an $f \in A$ such that $f(x) \neq 0$ and $f^q \operatorname{id}_M \in P$ for some $q \geq 0$.

Select f so that M_f is A_f-free. Then the epimorphism $\psi_f : F_f \longrightarrow M_f$ has a section $\varphi : M_f \longrightarrow F_f$. By Lemma 1.10.7, for some $r \geq 0$ we can lift $f^r \varphi$ to a homomorphism $\chi : M \longrightarrow F$. The equality $\psi_f \circ \varphi = \operatorname{id}_{M_f}$ implies $(\psi \circ \chi)_f = f^r \operatorname{id}_{M_f}$; in particular,

$$(\psi \circ \chi - f^r \operatorname{id}_M)_f (m_i) = 0$$

for a finite collection of generators m_i of M. Therefore, $f^t (\psi \circ \chi - f^r \operatorname{id}_M) = 0$ for some $t \geq 0$; hence $f^{r+t} \operatorname{id}_M = \psi \circ f^t \chi \in P$.

Now, select a finite subcover $\bigcup_{1 \leq i \leq k} D(f_i)$ in a cover of $\operatorname{Spec} A$ with big open sets; this is possible because $\operatorname{Spec} A$ is quasi-compact. Find q for which $f_i^q \operatorname{id}_M \in P$ for all i. Since $D(f_i^q) = D(f_i)$, it follows that $\{f_i^q\}_{1 \leq i \leq k}$ generates the unit ideal. The partition of unity $\sum_{1 \leq i \leq k} g_i f_i^q = 1$ implies that

$$\operatorname{id}_M = \left(\sum_{1 \leq i \leq k} g_i f_i^q \right) \operatorname{id}_M \in P. \qquad \square$$

1.10.9 Nakayama's Lemma

Now, we would like to establish that projective modules are locally free. First, we verify this for a stronger localization procedure.

The following simple but fundamental result is called *Nakayama's lemma*.

1.10.9a Lemma (Nakayama's Lemma) *Let A be a local ring, $\mathfrak{a} \subset A$ an ideal not equal to A, and M an A-module of finite type. If $M = \mathfrak{a}M$, then $M = \{0\}$.*

1.10.9a.i Examples Illustrating the Necessity of Finite Type Condition on the Modules

(a) Let A be a ring without zero divisors, M its field of fractions. Obviously, if $\mathfrak{a} \neq \{0\}$, then $\mathfrak{a}M = M$, yet $M \neq \{0\}$.

(b) Let A be the ring of germs of C^∞-functions in a vicinity of the origin of $\mathbb{R}$ and $\mathfrak{a}$ the ideal of functions that vanish at the origin. Let $M = \bigcap_{n \in \mathbb{Z}_+} \mathfrak{a}^n$ be the ideal of *flat* functions, i.e., the functions that vanish at the origin together with all their derivatives. It is not difficult to establish that $\mathfrak{a}M = M$: this follows from the fact that, for any flat function f and the coordinate function x, the quotient f/x, whose value at the origin is defined to be equal to 0, is a flat function.

Proof of Nakayama's Lemma Let $M \neq \{0\}$. Select a minimal finite family of generators $m_1, \ldots, m_2$ of M. Since $M = \mathfrak{a}M$, it follows that $m_1 = \sum_{1 \leq i \leq r} f_i m_i$ for $f_i \in \mathfrak{a}$, i.e., $(1 - f_1)m_1 = \sum_{i \geq 2} f_i m_i$. Since f_1 lies in a maximal ideal of A, it follows that $1 - f_1$ is invertible; therefore, m_1 can be linearly expressed in terms of $m_2, \ldots, m_r$. This contradicts the minimality of the family of generators. $\square$

1.10.9b Corollary *Let M be a module of finite type over a local ring A with maximal ideal $\mathfrak{p}$. Let the elements $\overline{m}_i = m_i \pmod{\mathfrak{p}M}$, where $i = 1, \ldots, r$, span $M/\mathfrak{p}M$ as a linear space over the field $A/\mathfrak{p}$. Then the m_i generate the A-module M. In particular, if A is Noetherian, the basis elements of the $A/\mathfrak{p}$-space $\mathfrak{p}/\mathfrak{p}^2$ generate the ideal $\mathfrak{p}$.*

Proof Let $M' = M/(Am_1 + \cdots + Am_r)$. Since $M = \mathfrak{p}M + Am_1 + \cdots + Am_r$, we see that $M' = \mathfrak{p}M'$, implying $M' = 0$. $\square$

1.10.10 Proposition *Any projective module M of finite type over a local ring A is free.*

Proof Let $\mathfrak{p}$ be the maximal ideal in A. Then $M/\mathfrak{p}M$ is a finite-dimensional vector space over $A/\mathfrak{p}$; let $\overline{m}_i = m_i \pmod{\mathfrak{p}M}$, where $i = 1, \ldots, r$, be a basis. By the above, the m_i constitute a family of generators of M. Let us show that M is free. Consider an epimorphism $F \longrightarrow M \longrightarrow 0$, where $F = A^r$ is a free module of rank r whose free generators are mapped into the $\{m_i\}_{i=1}^r$. Since M is projective, there exists a section $\varphi: M \longrightarrow F$ which induces an isomorphism $\overline{\varphi}: M/\mathfrak{p}M \longrightarrow F/\mathfrak{p}F$ because the dimensions of both spaces are equal to r.

Therefore, either $F = \varphi(M) + \mathfrak{p}F$, or $F/\varphi(M) = \mathfrak{p}(F/\varphi(M))$. By Nakayama's lemma, $F = \varphi(M)$; hence, φ is an isomorphism. $\square$

1.10.11   Proof of Theorem 1.10.4, Completion

1.10.11a Proposition *Any Noetherian projective A-module M over a Noetherian ring A is locally free.*

Proof Let $x \in \operatorname{Spec} A$ and $\mathfrak{p} \subset A$ the corresponding prime ideal. Since the module $M_{\mathfrak{p}} = A_{\mathfrak{p}} \otimes M$ is projective, and therefore free thanks to Proposition 1.10.10. Take an $A_{\mathfrak{p}}$-basis in $M_{\mathfrak{p}}$. Reducing the elements of the basis to the common denominator we may assume that they are of the form m_i/g, where $m_i \in M$ for $i = 1, \ldots, n$ and $g \in A$. Consider a homomorphism $\varphi: A_g^n \longrightarrow M_g$ sending the elements of a free basis of A_g^n to m_i/g; set $K = \operatorname{Ker} \varphi$ and $C = \operatorname{Coker} \varphi$.

Tensoring the exact sequence of A_g-modules

$$0 \longrightarrow K \longrightarrow A_g^n \longrightarrow M_g \longrightarrow C \longrightarrow 0 \tag{1.12}$$

by $A_{\mathfrak{p}}$, which is also the localization of A_g modulo $A_g \setminus \mathfrak{p}_g$, we get thanks to Theorem 1.10.6 an exact sequence of $A_{\mathfrak{p}}$-modules. Its middle arrow is an isomorphism, and therefore $A_{\mathfrak{p}} \otimes_{A_g} K = 0$ and $A_{\mathfrak{p}} \otimes_{A_g} C = 0$.

Let $k_1, \ldots, k_s$ and $c_1, \ldots, c_r$ be bases of the A_g-modules K and C, respectively. By Lemma 1.10.9, there exist $h_i, h'_j \in A_g \setminus \mathfrak{p}_g$ such that $h_i k_i = 0$ and $h'_j c_j = 0$ for all i and j. In particular,

$$
h = \left(\prod_{1 \leq i \leq s} h_i \right) \left(\prod_{1 \leq j \leq r} h'_j \right) \in A_g \setminus \mathfrak{p}_g
$$

and h annihilates K and C. Let $h = f/g^k$, where $f \in A \setminus \mathfrak{p}$. Then $f/1$ annihilates both K and C. Tensoring (1.12) by $(A_g)_{f/1}$ over A_g, and using the A_{fg}-module isomorphism $(M_g)_{f/1} \simeq M_{fg}$, we see that there exists an isomorphism $A_{fg}^n \simeq M_{fg}$, because $K_{f/1} = \{0\}$ and $C_{f/1} = \{0\}$. Since $fg(x) \neq 0$, we conclude that M is locally free at x. □

1.10.12 An Example of a Non-free Projective Module

Let A be the ring of real-valued continuous functions on $[0, 1]$ such that $f(0) = f(1)$, i.e., A may be viewed as the ring of functions on the circle S^1. The module of sections of the Möbius bundle over S^1 may be described as the A-module M of functions on $[0, 1]$ such that $f(0) = -f(1)$.

1.10.12a Theorem *M is not free, but* $M \oplus M \cong A \oplus A$.

Proof
(1) For any $f_1, f_2 \in M$, we have $f_1 f_2 (f_2) - f_2^2 (f_1) = 0$ and $f_1 f_2, f_2^2 \in A$; hence, any two elements of M are linearly dependent over A. This means that if M is free, then $M \cong A$.

But $M \not\cong Am$ for any $m \in M$ because m vanishes somewhere on $[0, 1]$ thanks to continuity, whereas M possesses elements that do not vanish at any prescribed point.

(2) The elements $f = (\sin \pi t, \ \cos \pi t)$ and $g = (-\cos \pi t, \ \sin \pi t)$ constitute a free basis of $M \oplus M$ because, for any $(m_1, m_2) \in M \oplus M$, the system of equations

$$
\begin{cases}
x \sin \pi t - y \cos \pi t = m_1, \\
x \cos \pi t + y \sin \pi t = m_2,
\end{cases}
$$

is uniquely solvable in A. □

1.11   The Normal Bundle and Regular Embeddings

1.11.1   Conormal Module

Let Y be a closed subscheme of an affine scheme $X = \operatorname{Spec} A$ determined by an ideal $\mathfrak{a}$. Then the $A/\mathfrak{a}$-module $\mathfrak{a}/\mathfrak{a}^2$ is called the *conormal module of Y with respect to the embedding* $Y \hookrightarrow X$, or simply the *conormal module* to Y and the family of vector spaces $N = \operatorname{Spec} S_{A/\mathfrak{a}}(\mathfrak{a}/\mathfrak{a}^2)$ is called the *normal family*.

The following geometric picture illustrates this definition: let $\mathfrak{a}$ be the ideal of functions on X that vanish on Y. Then $\mathfrak{a}^2$ is the ideal of functions whose zeroes on Y are of order ≥ 2, and $\mathfrak{a}/\mathfrak{a}^2$ is the module of linear parts of these functions in a neighborhood of Y. A *tangent vector to X at a point* $y \in Y$ determines a linear function on such linear parts. A *normal vector to Y at y* (in the absence of a natural metric) is a class of tangent vectors to X at $y \in Y$ modulo those that are tangent to Y, i.e., the ones that vanish on the linear parts of the functions (i.e., elements) of $\mathfrak{a}$. Therefore, in "sufficiently regular" cases, $\mathfrak{a}/\mathfrak{a}^2$ is (locally) the space dual to the space of vectors normal to Y. This explains the meaning of the term *conormal module*.

1.11.2   Regular Embeddings

The conormal module is, in general, neither free, nor projective, but it is both free and projective for a very important class of subschemes.

A sequence of elements $(f_1, \ldots, f_n)$ of a ring A is said to be *regular* (of length n) if, for all $i \geq n$, the element $f_i \mod (f_1, \ldots, f_{i-1})$ is not a zero divisor in $A/(f_1, \ldots, f_{i-1})$; it is convenient to assume that the empty sequence is regular of length 0 and generates the zero ideal.

A closed subscheme $Y \subset X = \operatorname{Spec} A$ is said to be *regularly embedded* or, more often, a *complete intersection* (of codimension n), if A contains a regular sequence of length n generating the defining ideal of Y.

The geometric meaning of the notion of complete intersection becomes totally transparent when we recall that we define Y by adding one of the equations $f_i = 0$ at a time. Thus, we get a decreasing sequence of subschemes

$$X \supset Y_1 \supset Y_2 \supset \cdots \supset Y_n = Y.$$

The complete intersection condition means that Y_i does not contain the whole support of any of the components of the incontractible primary decomposition of Y_{i-1}. In other words, each equation $f_i = 0$ should be "transverse" (in a very weak sense) to all these supports.

1.11.2i Proposition *If* $Y \hookrightarrow X$ *is a complete intersection, then its conormal module is free. In particular, the rank n of the conormal module does not depend on the choice of a regular sequence of generators of the ideal.*

The rank n of the conormal module is called the *codimension* of Y in X.

Proof Let $\mathfrak{a} = (f_1, \ldots, f_n) \subset A$, where $f_1, \ldots, f_n$ is a regular sequence. Obviously, the elements $\overline{f}_i = f_i \pmod{a}$ generate the $A/\mathfrak{a}$-module $\mathfrak{a}/\mathfrak{a}^2$. Therefore, it suffices to verify that they are linearly independent. This is done by induction on n:

First, let $n = 1$. Then $f_1 = f$, $\overline{g} = g \pmod{A_f}$. If $\overline{g}\overline{f} = 0$, then $gf = hf^2$ for some $h \in A$, and so $f(g - hf) = 0$; hence $g = hf$, because f is not a zero divisor in A. Therefore, $\overline{g} = 0$.

Let the result be already proved for a regular sequence $\{f_1, \ldots, f_{n-1}\}$. Assume that $\sum_{1 \le i \le n} \overline{g}_i\overline{f}_i = 0$ in $\mathfrak{a}/\mathfrak{a}^2$, where $\overline{g}_i = g_i \pmod{a}$. We may assume that $\sum_{1 \le i \le n} g_i f_i = 0$ in A: otherwise,

$$\sum_{1 \le i \le n} g_i f_i = \sum_{1 \le i \le n} u_i f_i, \quad \text{where } u_i \in \mathfrak{a},$$

and we can replace g_i with $g_i - u_i$ without affecting $\overline{g}_i$.

Since the class $\overline{f}_n$ is not a zero divisor in $A/(f_1, \ldots, f_{n-1})$, it follows that

$$g_n f_n + \sum_{1 \le i \le n-1} g_i f_i = 0 \implies g_n \in (f_1, \ldots, f_{n-1})$$

i.e., $g_n = \sum_{1 \le i \le n-1} h_i f_i$, whence

$$\sum_{1 \le i \le n-1} (g_i + h_i f_n) f_i = 0.$$

By the induction hypothesis, $g_i + h_i f_n \in (f_1, \ldots, f_{n-1})$ for $i = 1, \ldots, n-1$; hence, $g_i \in \mathfrak{a}$ for all i, i.e., $\overline{g}_i = 0$. $\square$

1.11.3   Locally Regularly Embedded Subschemes

Let $Y \hookrightarrow X$ be a subscheme. This subscheme is said to be *locally regularly embedded at* $y \in Y$ if there exists a neighborhood $D(f) \ni y$ such that $Y \cap D(f)$ is regularly embedded into $D(f)$. Obviously, $Y \cap D(f)$ is determined by an ideal $\mathfrak{a}_f \subset A_f$ and coincides with the fiber product $Y \times_X X_f$.

1.11.3a Statement *The normal family to any locally regularly embedded subscheme is a vector bundle.*

Indeed, $(\mathfrak{a}/\mathfrak{a}^2)_f = \mathfrak{a}_f/\mathfrak{a}_f^2$, so that the $A/\mathfrak{a}$-module $\mathfrak{a}/\mathfrak{a}^2$ is locally free for such a subscheme. $\square$

1.11.3b Remark It may well happen that a subscheme is regularly embedded locally, but not globally. The first example of such a situation was encountered in number theory.

Let $A \supset \mathbb{Z}$ be a ring of integer algebraic numbers of a field K. If the *class number* (for its definition, see, e.g., [BSh, Lan]) of K is greater than 1, then A has non-principal ideals $\mathfrak{a} \subset A$ (which are even prime). However, any such ideal, as is known, is "locally" principal. Therefore, $\mathfrak{a}$ determines a locally regularly embedded subscheme of codimension 1.

1.11.4   The Tangent and Cotangent Spaces

Let $x \in X$ be a closed point: for brevity, we will also denote by x the unique reduced subscheme with support at this point. Let $\mathfrak{m}_x$ be the maximal ideal corresponding to x. The discussion in Sect. 1.11.1 shows that $\mathfrak{m}_x/\mathfrak{m}_x^2$, the *co-normal module* to x, is an analogue of the cotangent space to X at x. This is the *Zariski cotangent space*. Its dual, $(\mathfrak{m}_x/\mathfrak{m}_x^2)^*$, is called the *tangent space* at x.

Closed points may or may not be locally regularly embedded.

For instance, all closed points of $\mathbb{A}^n = \operatorname{Spec} K[T_1, \ldots, T_n]$, where, for simplicity, K is assumed to be algebraically closed, correspond to the ideals of the form $(T - t_1, \ldots, T - t_n)$, where $t_i \in K$. The indicated sequence of generators of such an ideal (i.e., the sequence $T - t_1, \ldots, T - t_n$) is, obviously, a regular sequence.

To get examples of non-locally regularly embedded points, it suffices to consider the spectrum of a local Artinian ring which is not a field: all elements of its maximal ideal are nilpotents, and therefore there is no sequence of generators whose first element is not a zero divisor. More meaningful examples are provided by *hypersurfaces*, i.e., subschemes of the affine space $\mathbb{A}^n$ given by one equation.

1.11.5 Example Let $X \subset \mathbb{A}^n$ be a closed subscheme of an affine space over the field K through the origin 0; let the equation of X be $F = 0$, where $F = F_1 + F_2 + \cdots$ and F_i is a form of degree i in $T_1, \ldots, T_n$.

1.11.5a Statement *The point x is locally regularly embedded in X if and only if $F_1(x) = 0$.*

1.11.5 Corollary *Let $F(t_1, \ldots, t_n) = 0$, where $t_i \in K$. The point x defined by the ideal $(\ldots, T - t_i, \ldots)$ is locally regularly embedded in X if and only if there exists an i such that*

$$\frac{\partial F}{\partial T_i}(t_i, \ldots, t_n) \neq 0.$$

Proof Indeed, translate the origin to $(t_1, \ldots, t_n)$; then the linear part of F in a vicinity of the new origin is equal to

$$\sum \frac{\partial F}{\partial T_i}(t_1, \ldots, t_n)(T_i - t_i)$$

and it suffices to apply Statement 1.11.5a. $\qquad\square$

This differential criterion shows that locally regularly embedded points are exactly the ones which in classical algebraic geometry are called *nonsingular*.

Leaving a systematic theory of such points for the future, we confine ourselves here to general facts needed for studying Example 1.11.5.

Note, first of all, that it suffices to consider localized rings. More precisely, let $\mathfrak{p} = (T_1, \ldots, T_n) \subset A$, and $\overline{\mathfrak{p}} = \mathfrak{p} \bmod (F) \subset B$. The proof of Proposition 1.11.2i shows that the point x is locally regularly embedded into X if and only if the maximal ideal of the local ring $B_{\overline{\mathfrak{p}}}$ is generated by a regular sequence.

Observe that $B_{\overline{\mathfrak{p}}} = A_{\mathfrak{p}}/(F/1)$.

1.11.5c Lemma *Under the conditions of Example* 1.11.5, *if* $F_1 \neq 0$, *then the maximal ideal in* $B_{\overline{\mathfrak{p}}}$ *is generated by a regular sequence.*

Proof Indeed, making an invertible linear change of indeterminates we may assume that $F_1 = T_1$. Now, for any $G \in K[T_1, \ldots, T_n]$, let $\overline{G}$ denote the class $G/1 \bmod (F/1)$ in the ring $B_{\overline{\mathfrak{p}}}$.

The elements $T_2/1, \ldots, T_n/1$ and $F/1$ form a regular sequence in the ring $A_{\mathfrak{p}}$ because $F \equiv T_1 + a_2 T_1^2 + \cdots \bmod (T_2, \ldots, T_n)$, where $a_i \in K$. Corollary 2.9.7 (to be proved in Chap. 2) implies that $F/1, T_2/1, \ldots, T_n/1$ is also a regular sequence.

Therefore, $\overline{T}_2/1, \ldots, \overline{T}_n/1$ is a regular sequence in $B_{\overline{\mathfrak{p}}} = A_{\mathfrak{p}}/(F/1)$; clearly, this regular sequence generates a maximal ideal in $B_{\overline{\mathfrak{p}}}$. $\qquad\square$

1.11.6 The Converse to Lemma 1.11.5c

To prove this converse statement, observe that if the origin is locally regularly embedded in X, then the maximal ideal of the local ring $B_{\overline{\mathfrak{p}}}$ must be generated by a regular sequence. The condition $F_1 = 0$ means that $F/1$ belongs to the square of the maximal ideal in $A_{\mathfrak{p}}$. Therefore, it suffices to establish the following lemma.

1.11.6a Lemma *Let A be a local Noetherian ring, $\mathfrak{p} \subset A$ its maximal ideal generated by a regular sequence of length n. If $f \in \mathfrak{p}^2$ and f is regular, then the maximal ideal in the local ring $A/(f)$ cannot be generated by a regular sequence.*

Proof Suppose the maximal ideal in $A/(f)$ is generated by a maximal sequence $\overline{g}_1, \ldots, \overline{g}_k$, where $\overline{g}_i = g_i \bmod (f)$, and $g_i \in A$. Then $(f, g_1, \ldots, g_k)$ is a regular sequence in A that generates $\mathfrak{p}$. Since the length of any such sequence is equal to n (Prop. 1.11.2i), we should have $k = n - 1$. But since $f \in \mathfrak{p}^2$, the elements $(f, g_1, \ldots, g_k)$ span a subspace of dimension $\leq k = n - 1$ in the n-dimensional

$A/\mathfrak{p}$-space $\mathfrak{p}/\mathfrak{p}^2$. The contradiction obtained proves the lemma and completes the study of Example 1.11.5c.   □

1.12 Differentials

1.12.1 The Module of Universal Differentials

Let A and B be commutative rings and B an A-algebra. Let $\mu: B \otimes_A B \longrightarrow B$, where $\mu(b_1 \otimes b_2) := b_1 b_2$, be the multiplication in B. Set

$$I = I_{B/A} = \operatorname{Ker} \mu.$$

Clearly, I is an ideal in $B \otimes_A B$ and $(B \otimes_A B)/I \cong B$.

Consider the B-module[18] $\operatorname{Covect}_{B/A} = I/I^2$, called the *module of (relative) differentials* of the A-algebra B.

By Proposition 1.9.6, the ideal I defines the diagonal subscheme $\Delta_X \subset X \times_S X$, where $X = \operatorname{Spec} B$, $S = \operatorname{Spec} A$. According to the interpretation in Sect. 1.11.1, the module $\operatorname{Covect}_{B/A}$ represents the *conormal of the diagonal*.

In Differential Geometry, the normal bundle of the diagonal Δ is isomorphic to the tangent bundle of the manifold X itself. Indeed, by transporting a vector field along one of the fibers of the product $X \times X$ "parallel" to the diagonal; we get a vector field that is everywhere transverse to the diagonal, see Fig. 1.12. Therefore, $\operatorname{Covect}_{B/A}$ is a candidate for the role of the module "cotangent" to X along the fibers of the morphisms $X \longrightarrow S$.

On the other hand, in the interpretation of nilpotents given in Sect. 1.5 as an analog of the elements of the "tangent module" to X (over S), we have already considered the B-module $D_{B/A}$ of *derivations* of the A-algebra B (a *vector field on*

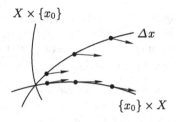

$X \times \{x_0\}$

Δx

$\{x_0\} \times X$

Fig. 1.12

---

[18]In Differential and Algebraic Geometries we sometimes need not only symmetric powers of differential forms, e.g., as in metrics, but also the **exterior powers**, and in **this** context the module of differentials is denoted by $\Omega^1_{B/A}$. For details, see [MaG, SoS].

X, i.e., a section of the tangent bundle, can be naturally interpreted as a derivation of the ring of functions on X).

1.12.1a Lemma
(1) The map d is an A-derivation, and $d(\varphi(a)) = 0$ for any $a \in A$, where $\varphi: A \longrightarrow B$ is the morphism that defines the A-algebra structure.
    (2) Let $\{b_i \mid i \in I\}$ be a set of generators of the A-algebra B.
    Then $\{db_i \mid i \in I\}$ is a set of generators of the B-module $\text{Covect}_{B/A}$.

Proof
(1) This is subject to a straightforward verification. For example, to prove the Leibniz rule $(d(b_1 b_2) = (db_1)b_2 - b_1(db_2))$, use the fact that the multiplication in by b in $\text{Covect}_{B/A}$ is induced by multiplication by $b \otimes 1$ or $1 \otimes b$ in $I_{B/A}$, and observe that

$$b_1 b_2 \otimes 1 - 1 \otimes b_1 b_2 = b_1 \otimes 1(b_2 \otimes 1 - 1 \otimes b_2) + (b_1 \otimes 1 - 1 \otimes b_1)1 \otimes b_2.$$

(2) Notice that

$$\sum_i b_i \otimes b_i' \in I_{B/A} \iff \sum_i b_i b_i' = 0 \iff$$

$$\sum_i b_i \otimes b_i' = \sum_i b_i \otimes b_i' - \sum_i b_i b_i' \otimes 1 = \sum_i b_i \otimes 1(1 \otimes b_i' - b_i' \otimes 1).$$

This implies that, as a B-module, $\text{Covect}_{B/A}$ is generated by all the elements db with $b \in B$. Since d is a derivation that vanishes on the image of A, this easily implies the desired conclusion. $\qquad\qquad\qquad\qquad\qquad\qquad\qquad\qquad\qquad\qquad\qquad\qquad\qquad\qquad\square$

1.12.1b Example Let $B = A[T_1, \ldots, T_n]$. Then $\text{Covect}_{B/A}$ is the free B-module freely generated by the elements dT_i.

1.12.1c Remark The higher-order "differential neighborhoods of the diagonal" are represented by the schemes $\text{Spec}(B \otimes_A B)/I^n_{B/A}$. They replace the spaces of *jets* of germs of the diagonal considered in Differential Geometry.

Define a map $d = d_{B/A}: B \longrightarrow \text{Covect}_{B/A}$ by setting

$$d(b) = (b \otimes 1 - 1 \otimes b) \pmod{I^2_{B/A}}.$$

In Differential Geometry, the tangent and cotangent bundles are dual to each other. In the algebraic setting (over finite fields), this is not the case, generally: only "a half" of the duality is preserved:

$$D_{B/A} = \text{Hom}_B(\text{Covect}_{B/A}, B).$$

Therefore, $D_{B/A}$ is recovered from Covect$_{B/A}$, but not the other way round. This explains the advantage of differentials as compared with derivatives.[19]

1.12.2 Proposition *Let* $d': B \longrightarrow M$ *be a derivation of* B *into a* B-*module* M *that vanishes on the image of* A. *Then there exists a unique* B-*module homomorphism* $\psi: \mathrm{Covect}_{B/A} \longrightarrow M$ *such that* $d' = \psi \circ d_{B/A}$.

(Applying this result to $M = B$, we get (1.12.1)).

Proof The *uniqueness* of ψ follows immediately from the fact that $d'b = \psi(db)$ for all $b \in B$, so that ψ is uniquely determined on the family of generators of Covect$_{B/A}$.

To prove the *existence*, let us first define a group homomorphism

$$\chi: B \otimes_A B \longrightarrow M, \qquad \chi(b \otimes b') = bd'b'.$$

This homomorphism vanishes on $I^2_{B/A}$. Indeed: first, notice that χ is a B-module homomorphism with respect to the B-action $b \longmapsto b \otimes 1$ on $B \otimes_A B$. Furthermore, as shown above, the elements $b \otimes 1 - 1 \otimes b$ generate the B-module $I_{B/A}$, and therefore the products

$$(b_1 \otimes 1 - 1 \otimes b_1)(b_2 \otimes 1 - 1 \otimes b_2)$$

---

[19]Grothendieck showed that one can, however, define *differential operators of order* $\leq k$ for any (commutative) ring R over K or $\mathbb{Z}$ as the K- or $\mathbb{Z}$-linear maps of R-modules $D : M \longrightarrow N$ such that

$$[l_{r_0}, [l_{r_1}, \ldots [l_{r_k}, D] \ldots]] = 0 \quad \text{for any } r_0, r_1, \ldots r_k \in R,$$

where l_r denotes the operator of left multiplication by r in M and in N. Denote by Diff$_k(M, N)$ the R-module of differential operators of degree $\leq k$; set

$$\mathrm{Diff}(M, N) = \varinjlim \mathrm{Diff}_k(M, N).$$

Define the R-module of *symbols of differential operators* to be the graded space associated with Diff(M, N)

$$\mathrm{Smbl}(M, N) = \bigoplus \mathrm{Smbl}_k(M, N),$$

where

$$\mathrm{Smbl}_k(M, N) = \mathrm{Diff}_k(M, N) / \mathrm{Diff}_{k-1}(M, N).$$

If $M = N$, then, clearly, the space Diff $(M) := \mathrm{Diff}(M, M)$ can be viewed as an associative algebra with respect to the product-juxtaposition and a Lie algebra, denoted by $\mathfrak{diff}(M)$, with respect to the commutator of operators. The space Smbl(M) can be viewed as a commutative R-algebra with respect to the product-juxtaposition. The commutator in $\mathfrak{diff}(M)$ induces a Lie algebra structure in Smbl(M) (the *Poisson bracket*); this Lie algebra is called the *Poisson algebra* in 2 dim M indeterminates over R.

generate the B-module $I_{B/A}^2$. Therefore, it suffices to verify that χ vanishes on such products. This is straightforward; hence, we see that χ induces a map $\varphi: I/I^2 \longrightarrow M$. We get

$$\varphi(d\,b) = \varphi(b \otimes 1 - 1 \otimes b) = d'b,$$

completing the proof. □

1.12.3 Conormal Bundles and Sheaves

Now, let $i: Y \hookrightarrow X$ be a closed embedding of schemes. In differential-geometric models, the restriction of the tangent bundle of X to Y contains, under certain regularity conditions, the tangent bundle of Y, and the corresponding quotient bundle is the *normal bundle* of Y. We would like to determine to what extent a similar statement is true for schemes.

Let us translate the problem into algebraic language.

Let B be an A-algebra, $\mathfrak{b} \subset B$ an ideal. Then $\overline{B} = B/\mathfrak{b}$ is also an A-algebra, and the relative (over Spec A) cotangent sheaves on Spec B and Spec $\overline{B}$ are represented by the modules $\mathrm{Covect}_{B/A}$ and $\mathrm{Covect}_{\overline{B}/A}$, respectively. On the other hand, the *conormal sheaf* to the embedding Spec $\overline{B} \longrightarrow$ Spec B is represented by the $B/\mathfrak{b}$-module $\mathfrak{b}/\mathfrak{b}^2$.

An analogue of the classical situation is given by the following proposition. Let

$$\delta\overline{e} = 1 \otimes_B d_{B/A}(e) \quad \text{for } \overline{e} \in \mathfrak{b}/\mathfrak{b}^2 \text{ be represented by an element } e \in \mathfrak{b}. \tag{1.13}$$

Since the map $d': B \longrightarrow \mathrm{Covect}_{\overline{B}/A}$ given by $d'f = d_{\overline{B}/A}(f \bmod \mathfrak{b})$ is an A-derivation, it factorizes through a uniquely determined B-module homomorphism $\mathrm{Covect}_{B/A} \longrightarrow \mathrm{Covect}_{\overline{B}/A}$; since the target is annihilated by multiplication by $\mathfrak{b}$, this homomorphism determines a $B/\mathfrak{b}$-module morphism

$$u: B/\mathfrak{b} \otimes_B \mathrm{Covect}_{B/A} \longrightarrow \mathrm{Covect}_{\overline{B}/A}.$$

1.12.3a Proposition *There exists an exact sequence of $B/\mathfrak{b}$-modules*

$$\mathfrak{b}/\mathfrak{b}^2 \overset{\delta}{\longrightarrow} B/\mathfrak{b} \otimes_B \mathrm{Covect}_{B/A} \overset{u}{\longrightarrow} \mathrm{Covect}_{\overline{B}/A} \longrightarrow 0. \tag{1.14}$$

Proof If $\overline{e} = 0$, i.e., $e \in \mathfrak{b}^2$, then $de \in \mathfrak{b}d_{B/A}\mathfrak{b}$, so that $1_{\overline{B}} \otimes d_{B/A}(e) = 0$. Hence, $\delta(\overline{e})$ does not depend on the choice of e. It is obvious that δ is a group homomorphism, and the compatibility with the $B/\mathfrak{b}$-action follows from the fact that

$$\delta(\overline{f}\overline{e}) = 1 \otimes_B d(fe) = 1 \otimes_B (edf + fde) = \overline{f} \otimes_B de = \overline{f}\delta(\overline{e})$$

for any $\overline{f} = f \bmod \mathfrak{b}$.

Further, it is clear that $u(1 \otimes_B d_{B/A} f) = d_{\bar{B}/A}(f \bmod b)$ and, therefore, u is an epimorphism.

It is easy to see that $u \circ \delta = 0$:

$$u \circ \delta(\bar{e}) = u(1 \otimes de) = d(e \bmod b) = 0.$$

Let us verify the exactness in the middle term of the sequence (1.14). Construct a homomorphism

$$v \colon \mathrm{Covect}_{B/A} \longrightarrow \bar{B} \otimes_B \mathrm{Covect}_{B/A} / \mathrm{Im}\,\delta$$

such that u and v are inverses to one another. For this, define first a derivation

$$d' \colon (\bar{B} \to \bar{B} \otimes_B \mathrm{Covect}_{B/A}) / \mathrm{Im}\,\delta$$

by setting

$$d'(\bar{f}) = 1 \otimes_B d_{B/A}(f) \bmod \mathrm{Im}\,\delta \quad \text{for any } \bar{f} = f \bmod b.$$

The result does not depend on the choice of a representative of $\bar{f}$ because

$$1 \otimes d_{B/A}(e) \in \mathrm{Im}\,\delta \quad \text{for any } e \in b.$$

This derivation determines the homomorphism v. Since

$$(u \circ v)(df) = df \quad \text{and} \quad (v \circ u)(1 \otimes df \bmod \mathrm{Im}\,\delta) = 1 \otimes df \bmod \mathrm{Im}\,\delta,$$

it follows that v and u are inverses to one another on certain of our modules, as we needed to show.   □

1.12.3b Remark The difference of the above constructions from similar ones in differential geometry is crucial: It may well happen that $\mathrm{Ker}\,\delta \neq 0$ even if the subscheme $Y \hookrightarrow X$ is regularly embedded. For example, let $X = \mathrm{Spec}\,\mathbb{Z}$ and $Y = \mathrm{Spec}\,\mathbb{Z}/p\mathbb{Z}$, where p is a prime, $S = \mathrm{Spec}\,\mathbb{Z}$. Then $\mathrm{Covect}_{X/S} = 0$ and $\mathrm{Covect}_{Y/S} = 0$, whereas $(p)/(p)^2$ is a one-dimensional linear space over $\mathbb{Z}/p\mathbb{Z}$.

Informally speaking,[20]

it is impossible to differentiate in the "arithmetic direction".

---

[20]However, A. Buium worked out and found remarkably strong applications of the idea that "the derivative of an integer a with respect to a prime p is given by the Fermat quotient $\frac{a-a^p}{p}$", see his monograph [Bu] and a brief survey [MaN].

1.13 Digression: Serre's Problem and Seshadri's Theorem

Serre posed the following problem: *Over an n-dimensional affine space, are there non-trivial vector bundles?*

In other words, is the following statement true?
Any projective Noetherian module over $K[T_1, \ldots, T_n]$, where K is a field, is free.

For $n = 1$, the ring $K[T]$ is an integral principal ideal ring. Therefore, any Noetherian torsion-free $K[T]$-module (in particular, any projective Noetherian module) is free, see [Lang].

For $n = 2$, there are no non-trivial bundles, either. This theorem is due to Seshadri; this section is devoted to its proof.

For $n \geq 3$, the answer to Serre's question remained unknown for some time.[21] The problem is very attractive and has all features of a classical one: it is natural, pertains to fundamental objects, and is difficult. In any case, for 10 years after its formulation no essentially new results on modules over polynomial rings were obtained apart from Seshadri's theorem and the following fact due to Serre himself.

1.13.1 Theorem *Let P be a projective Noetherian $K[T]$-module, $T = (T_1, \ldots, T_n)$. Then there exists a free Noetherian module F such that $P \oplus F$ is free.*
In the language of topologists, vector bundles over affine spaces are stably free.

Proof follows easily from Hilbert's syzygy theorem, see Sect. 2.13.6.

Therefore, we confine ourselves to Seshadri's theorem. It is applicable to a class of rings containing, e.g., $K[T_1, T_2]$ and $\mathbb{Z}[T]$.

1.13.2 Theorem (Seshadri's Theorem) *Let A be an integral principal ideal ring. Then any projective Noetherian $A[T]$-module P is free.*

The proof is split into a series of lemmas. Its driving force is the simple remark that if A is a field, the statement is true. From A one can cook a field in two ways: pass from A to its field of fractions K, or pass to the quotient field $k = A/(p)$, where p is any prime element. Accordingly, the modules $K[T] \otimes_{A[T]} P$ and $k[T] \otimes_{A[T]} P$ turn out to be free. Now let us use these circumstances in turn.

1.13.3 Lemma *There exists an exact sequence of $A[T]$-modules*

$$0 \longrightarrow F \longrightarrow P \longrightarrow P/F \longrightarrow 0 \tag{1.15}$$

with the following properties:

(a) *F is a maximal $A[T]$-free submodule of P;*
(b) *$(\mathrm{Ann}(P/F)) \cap A \neq \{0\}$.*

---

[21] The affirmative answer ("Any projective Noetherian $K[T]$-module is free") is due to A. Suslin and D. Quillen [Q, Su, VSu]. L. Vasershtein later gave a simpler and much shorter proof of the theorem which can be found in Lang's book [Lang].

Proof Let $m'_1, \ldots, m'_r$ be a free $K[T]$-basis of the module $K[T] \otimes P$. There exists an element $0 \neq f \in A$ such that $m_i = fm' \in P_i \longleftrightarrow K[T] \otimes P$. The submodule $F' \subset P$ generated by the elements m_i, where $i = 1, \ldots, r$, is free. On the other hand, any element of a finite fixed family of generators of the module P is represented in $K[T] \otimes P$ as a linear combination $\sum_i F_{ij}(t)m_i$, where $F_{ij}(T) \in K[T]$. The common denominator of all coefficients of all polynomials $F_{ij}(T) \in A$ annihilates P/F'. Now, for the role of F we may take a maximal free submodule in P containing F': it exists thanks to the Noetherian property. Clearly, $\text{Ann}(P/F) \supset \text{Ann}(P/F')$, so $\text{Ann}(P/F) \cap A \neq \{0\}$.   □

We retain the notation of Lemma 1.13.3 and *intend to deduce a contradiction from the assumption that $F \neq P$.* In this case, $\text{Ann}(P/F) \cap A = (f) \subset A$, where f is non-invertible (because A is a principal ideal ring). Let p be a prime element of A dividing f. Set $k := A/(p)$ and tensor the exact sequence (1.15) by $k[T]$ over $A[T]$, having set $\overline{F} := F/pF = k[T] \otimes_{A[T]} F$ and so on:

$$\overline{F} \xrightarrow{\ i\ } \overline{P} \longrightarrow P/F \longrightarrow 0.$$

Let $\overline{F}_1 = \text{Ker}\, i$, and $\overline{F}_2 = \text{Im}\, i$. Since $\overline{P}$ is projective over $k[T]$, it follows that $\overline{F}_2$ is torsion-free, and hence free. Therefore, $\overline{F}_1$ is also free and is singled out in $\overline{F}$ as a direct summand, so one has a split sequence of free $k[T]$-modules:

$$0 \longrightarrow \overline{F}_1 \xrightarrow{\ i\ } \overline{F} \xrightarrow{\ j\ } \overline{F}_2 \longrightarrow 0. \tag{1.16}$$

1.13.4 Lemma $\overline{F}_1 \neq 0$.

Proof Indeed, $j(\overline{F}_1) = (pP \cap F)/pF$. Let $f = pg$. Since $g \notin (\text{Ann}(P/F)) \cap A$, the fact that $gP \not\subset F$ implies that $pgP \not\subset pF$, because P is torsion-free.

But $pgP = fP \subset pP \cap F$, hence the more so, $pP \cap F \not\subset pF$.   □

The last step requires some additional arguments.

1.13.5 Lemma *There exists a free $A[T]$-submodule $F_1 \subset F$ with a free direct complement and such that $k[T] \otimes F_1 = j(\overline{F}_1)$.*

Informally speaking, the sequence (1.16) can be lifted to a split exact sequence of free $A[T]$-modules.

1.13.6   Completion of the Proof of Theorem 1.13.2

Let $F_1 \subset F$ be the submodule whose existence is claimed in Lemma 1.13.5, $F_2 \subset F$ its free direct complement. Since $F_1/pF_1 = \text{Ker}\, i$, it follows that all elements of $F_1 \subset F \subset P$ are divisible by p inside P. Set

$$F'_1 = \{m \in P \mid pm \in F_1\}.$$

Clearly, F_1' is free (the multiplication by p establishes an isomorphism $F_1' \simeq F_1$) and is strictly bigger than F_1 (by Lemma 1.13.4). Hence, the module $F' = F_1' \oplus F_2 \subset P$ is free and contains F as a proper submodule, contradicting the maximality of F and completing the proof of Seshadri's theorem. $\qquad\square$

1.13.7 Proof of Lemma 1.13.5

Any automorphism φ of the module F induces an automorphisms φ of the module $\bar{F}$. We need the following auxiliary statement:

1.13.7a Lemma *The map* $\mathrm{SL}(n, A[T]) \longrightarrow \mathrm{SL}(n, k[T])$ *given by* $\varphi \longmapsto \bar{\varphi}$ *is surjective.*

Proof We use a classical result on reduction of a matrix with elements over the Euclidean ring $k[T]$ to diagonal form by "admissible transformations". For this result, see the book [vdW, § 85], where it is given in terms of bases. To formulate it, denote by I_n the unit $(n \times n)$-matrix over $k[T]$, by $I_{(ij)}$ the matrix obtained from I by transposition of the i-th and j-th rows, and by E_{ij} the *matrix unit*, i.e., the matrix with a 1 in the (ij)-the slot and 0 elsewhere.

The proof of the *theorem on elementary divisors* in the book [vdW] shows, in particular, that in a fixed basis of F any automorphism with determinant 1 can be represented as a product of matrices of the following types:

(1) $I + f E_{ij}, f \in k[T]$;
(2) $I_{(ij)}$;
(3) diagonal matrices with elements in k and with determinant 1.

The matrices of the first two types can obviously be lifted to elements of $\mathrm{SL}(n, A[T])$. The matrices of the third type can be factorized in a product of diagonal matrices with determinant 1 and such that only two of their diagonal elements are $\neq 1$. Therefore, we have reduced the problem to lifting matrices of the form $\begin{pmatrix} \bar{f} & 0 \\ 0 & \bar{f}^{-1} \end{pmatrix} \in \mathrm{SL}(2, k)$ to matrices in $\mathrm{SL}(2, A)$.

This can be done in a completely elementary way. Thanks to the Chinese remainder theorem, one can choose relatively prime elements $f, g \in A$ such that $\bar{f} \equiv f \mod (p)$ and $\bar{f}^{-1} \equiv g \mod (p)$. Now, we have $fg = 1 + ph$. In A, we solve the equation $fx + gy = h$; then

$$(f - py)(g - px) \equiv 1 + p^2 xy,$$

so the matrix $\begin{pmatrix} f - py & px \\ py & g - px \end{pmatrix}$ is the solution desired. $\qquad\square$

Now, let us return to the proof of Lemma 1.13.5. Select a free $A[T]$-basis $(m_i)_{i \in I}$ of the module F; its reduction modulo p is a free $k[T]$-basis $(\overline{m}_i)_{i \in I}$ of the module $\overline{F}$. Further, select a free $k[T]$-basis $(\overline{n}_i)_{i \in I}$ of the module $\overline{F}$ compatible with the split sequence (1.16), in the sense that the first rk $\overline{F}_1$ of its elements constitute a basis of $\overline{F}_1$. We may assume that the matrix $\overline{M} \in \mathrm{GL}(n, k[T])$ sending the set $(\overline{m}_i)_{i \in I}$ into the set $(\overline{n}_i)_{i \in I}$ belongs to $\mathrm{SL}(n, k[T])$: if not, it suffices to replace $\overline{n}_1$ with $(\det \overline{M})^{-1} \overline{n}_1$. Let us now lift $\overline{M}$ to $M \in \mathrm{SL}(n, A[T])$ and let $(n_i)_{i \in I}$ be an $A[T]$-basis of the module F. Further, let F_1 be the submodule of F generated by the first rk $\overline{F}_1$ elements of the basis $(n_i)_{i \in I}$, and F_2 the submodule generated by the remaining elements. The construction of these submodules shows that they satisfy Lemma 1.13.5, completing the proof. $\square$

1.14   Digression: ζ-function of a Ring

1.14.1   An Overview

The rings of finite type over a field are called *geometric rings*, those over $\mathbb{Z}$ *arithmetic rings*. These two classes of rings have a nonzero intersection: the rings of finite type over finite fields. Such rings (and their spectra) enjoy a blend of arithmetic and geometric properties as demonstrated by A. Weil in his famous conjectures on ζ-functions, proved by P. Deligne.

Here we introduce ζ-functions of arithmetic rings and indicate their simplest properties. A motivation for introducing the ζ-function: the closed points x in the spectrum of an arithmetic ring have a natural "norm" $N(x)$, equal to the number of elements in the finite field $k(x)$, and the number of points of given norm is finite. It is natural to expect that directly counting such points we get an interesting invariant of the ring.

Let A be an arithmetic ring. Let $n(p^a)$ be the number of closed points $x \in \mathrm{Spec}\, A$ such that $N(x) = p^a$, and $\nu(p^a)$ be the number of geometric $\mathbb{F}_{p^a}$-points of A.

1.14.2 Lemma   *The numbers $\nu(p^a)$ and $n(p^a)$ are finite and related as follows:*

$$\nu(p^a) = \sum_{b \mid a} b n(p^b). \tag{1.17}$$

Proof Every geometric $\mathbb{F}_{p^a}$-point of A is by definition a homomorphism $A \longrightarrow \mathbb{F}_{p^a}$. Consider all geometric points with the same center x; then $\mathfrak{p}_x \subset A$ is the kernel of the corresponding homomorphism and its image coincides with the unique subfield $\mathbb{F}_{p^b} \hookrightarrow \mathbb{F}_{p^a}$, where $p^b = N(x)$. There are exactly b homomorphisms with fixed kernel and image, because $\mathbb{F}_{p^b} / \mathbb{F}_{p^a}$ is a Galois extension of degree b. Therefore,

$$\nu(p^a) = \sum_{b \mid a} b \left(\sum_{\{x \mid N(x) = p^b\}} 1 \right) = \sum_{b \mid a} b n(p^b).$$

(This equality has an obvious meaning even if it is not known that $v(p^a)$ and $n(p^b)$ are finite.)

In particular, $n(p^a) \leq v(p^a)$, and it suffices to prove that $v(p^a)$ is finite. We identify Spec A with a closed subset in Spec $\mathbb{Z}[T_1, \ldots, T_n]$; then $N(x)$ does not depend on whether we consider x as belonging to Spec A or to Spec $\mathbb{Z}[T_1, \ldots, T_n]$. Therefore, with obvious notation,

$$v(p^a) \leq v_{\mathbb{Z}[T_1, \ldots, T_n]}(p^a) = p^{na}$$

where, clearly, p^{na} is just the number of geometric points of the n-dimensional affine space over a field with p^a elements. $\square$

1.14.3 ζ-functions of Arithmetic Rings

Define the ζ-function of any arithmetic ring A first formally, by setting

$$\zeta_A(s) = \prod_{x \in \mathrm{Spm}\, A} \frac{1}{1 - N(x)^{-s}}.$$

Clearly, for $A = \mathbb{Z}$, we get the usual *Euler function* (*Euler product*)

$$\zeta_{\mathbb{Z}}(s) = \prod_p \frac{1}{1 - p^{-s}} = \zeta(s).$$

The relation of the ζ-function with $n(p^a)$ and $v(p^a)$ is given by the following obvious identity

$$\zeta_A(s) = \prod_p \prod_{1 \leq a < \infty} \frac{1}{(1 - p^{-as})^{n(p^a)}} = \prod_p \zeta_{A/pA}(s) \tag{1.18}$$

and another, a trifle less obvious one, in terms of a *Dirichlet series*[22]:

$$\ln \zeta_A(s) = \sum_p \sum_a v(p^a) \frac{1}{ap^{as}}. \tag{1.19}$$

---

[22]Recall that a *Dirichlet series* is any series of the form $\sum_{n=1}^{\infty} \frac{a_n}{n^s}$, where $s \in \mathbb{C}$, and $a_n \in \mathbb{C}$ for all n.

Proof of (1.19) (use Lemma 1.14.2)

$$\ln \zeta_A(s) = -\sum_p \sum_a \ln(1 - p^{-bs}) n(p^b) = -\sum_p \sum_{b \in \mathbb{N}} \sum_{k \in \mathbb{N}} n(p^b) \frac{1}{kp^{bks}}$$

$$= \sum_p \sum_{a \in \mathbb{N}} \sum_{b|a} \frac{b}{ap^{as}} n(p^b) = \sum_p \sum_a v(p^a) \frac{1}{a^s p^{as}}.$$

$$(1.20)$$

$$\square$$

Therefore, the calculation of the ζ-function is equivalent to that of $n(p^a)$ or $v(p^a)$ for all p, a. $\square$

Relation (1.18) shows that $\zeta_A(s)$ factorizes into the product of ζ-functions of rings of finite type over finite fields. This, however, does not mean in the least that the study of ζ-functions reduces to the case of such rings; the Riemann ζ-function shows how nontrivial the behavior of the global ζ-function can be even for the simplest local factors.

Even the individual p-factors can have sufficiently complicated structure if A is nontrivial.

A part of Weil's conjectures proved by Dwork[23] shows, however, that $\zeta(s)$ is a rational function in p^{-s} for any ring A of finite type over a field of characteristic p. For such rings, it is convenient to change the variable by setting $p^{-s} = t$ and set $\zeta_A(s) = Z_A(t)$. Then formula (1.19) shows that

$$\ln Z_A(t) = \sum_a \frac{v(p^a)t^a}{a},$$

or

$$\frac{Z_A'(t)}{Z_A(t)} = \sum_a v(p^a) t^{a-1}.$$

In particular, the rationality of $Z_A(t)$ establishes that the sequence $v_a = v(p^a)$ satisfies a recurrence relation of the form

$$v_{a+n} = \sum_{0 \le i \le n-1} r_i v_{a+i}$$

for sufficiently large a, with some fixed constants n and r_i. Since the v_a are the numbers of solutions of a system of equations with values in finite fields of growing degree, the statement on rationality has a direct arithmetic meaning.

---

[23] See [Kz]. By early 1970s Weil's conjectures (and even more difficult statements) were proved by Grothendieck and Deligne, see [SGA4, SGA4 1/2, Dan].

1.14.4 Frobenius Morphism

In any study of the ζ-function of a ring A over a field k of characteristic p the following circumstance is of fundamental importance: $\nu(p^k)$ can be viewed as the number of fixed points of a power of a certain map F acting on the set of geometric points of A.

The *Frobenius morphism* $F : A \longrightarrow A$ is the map $g \longmapsto g^p$ for any $g \in A$, where $p = \mathrm{Char}\, k$.

The same term—*Frobenius morphism*—is applied to the corresponding morphism of spectra, ${}^a F$, to their powers, F^n and $({}^a F)^n$, and to the maps induced by these maps on some other objects. In particular, let $\overline{\mathbb{F}}_p$ be the algebraic closure of the Galois field of characteristic p. Then F induces a map ${}^a F : A(\overline{\mathbb{F}}_p) \longrightarrow A(\overline{\mathbb{F}}_p)$ of the $\overline{\mathbb{F}}_p$-points of A into itself.

1.14.4a Proposition $A(\mathbb{F}_{p^b})$ *coincides with the set of fixed points of* F^b.

Proof Let $\varphi \in A(\overline{\mathbb{F}}_p)$ and let φ be represented by a homomorphism $\varphi : A \longrightarrow \overline{\mathbb{F}}_p$; let $F^b(\varphi)$ be represented by the homomorphism $f \longmapsto \varphi(f)^{p^b}$ for any $f \in A$. The condition $\varphi \in A(\mathbb{F}_{p^b})$ means that $\mathrm{Im}\, \varphi \subset \mathbb{F}_{p^b} \subset \overline{\mathbb{F}}_{p^b}$, i.e., $\varphi(f)^{p^b} = \varphi(f)$ for all f. Therefore, ${}^a F \varphi = \varphi$.

The converse statement follows from the Galois theory: $\mathbb{F}_{p^b}$ is the field of invariants for F^b. $\qquad\square$

1.14.5 Lefschetz Formula

If F is an endomorphism acting on a compact topological space V, then the number $\nu(F)$ of its fixed points (appropriately defined) satisfies the following famous *Lefschetz formula*:

$$\nu(F) = \sum_{0 \leq i \leq \dim V} (-1)^i \,\mathrm{tr}\, F|_{H^i(V)},$$

where the summands are the traces of the linear operators induced by F on the cohomology spaces of V with complex coefficients.

The role of compact topological spaces V is played in our setting by smooth projective schemes. For them, the essential part of Weil's conjectures states that the numbers $\nu(p^a)$ are always expressed by Lefschetz type formulas. So far, we have dealt with Euler products and Dirichlet series purely formally. Now, let us study their convergence a little.

Let A be an arithmetic ring, $\{x_i\}_{i \in I}$ the set of generic points of its irreducible components. Define the *dimension* of A setting

$$\dim A = \begin{cases} \max_i(\mathrm{tr}\ \deg\ k(x_i)) + 1, & \text{if } \mathbb{Z} \subset A, \\ \max_i(\mathrm{tr}\ \deg\ k(x_i)), & \text{if } \mathrm{Char}\, A > 0. \end{cases} \tag{1.21}$$

(The *transcendence degree* tr deg is calculated over the prime subfield of $k(x_i)$. The dimension thus introduced was considered already by Kronecker.)

1.14.5a Theorem *The Euler product* $\prod_{x \in \mathrm{Spm}A} \dfrac{1}{1 - N(x)^{-s}}$ *converges absolutely for* Re $s > \dim A$.

Proof We will verify the theorem by consecutively extending the class of rings considered. We assume that for the Riemann ζ-function $\zeta_{\mathbb{Z}}(s)$ the convergence of the product is known.

(a) Let $A = \mathbb{F}_p[T_1, \ldots, T_n]$. One can verify directly that expression (1.19) converges absolutely to $\ln(1 - p^{n-s})^{-1}$ for Re $s > n = \dim A$ because $v(p^a) = p^{an}$. This implies that under the same conditions the Euler product for A converges absolutely to $\dfrac{1}{1 - p^{n-s}}$.

(b) Let A be a ring without zero divisors and of finite type over $\mathbb{F}_p$. Let us apply Noether's normalization theorem, see Theorem 1.5.3f, and find a polynomial subring $B = \mathbb{F}_p[T_1, \ldots, T_n] \subset A$ such that A is a B-module with finitely many generators. There exists a constant d such that, over every geometric $\overline{\mathbb{F}}_p$-point of B, there are no more than d geometric points of A.

Indeed, let a homomorphism $B \longrightarrow \overline{\mathbb{F}}_p$ be given. To extend it to $A \longrightarrow \overline{\mathbb{F}}_p$, we have to define in $\overline{\mathbb{F}}_p$ the images of a finite number of generators of B over A, each of which is a root of an integer equation with coefficients in A. The images of these coefficients are already defined, and therefore the roots of equations are determined in finitely many ways.

This implies that $v_A(p^a) \le \alpha v_B(p^a) = \alpha p^{na}$, and therefore $\zeta_A(s)$ converges absolutely for Re $s > n = \dim A$, as above. Moreover, in this domain, we have

$$|\ln \zeta_A(s)| \le \alpha \ln(1 - p^{n-\sigma})^{-1}, \text{ where } \sigma = \text{Re } s.$$

(c) Let A be an arbitrary ring of finite type over $\mathbb{F}_p$. Let the $\mathfrak{p}_i \subset A$ be all minimal prime ideals of A, and $A_i = A/\mathfrak{p}_i$. Every geometric point of Spec A belongs to an irreducible component, and therefore

$$v_A(p^a) \le \sum_i v_{A_i}(p^a),$$

so the Euler product for A converges if Re $s > \max_i \dim A_i$ and in this domain satisfies

$$|\ln \zeta_A(s)| \leq \sum_i \alpha_i \ln(1 - p^{n_i - \sigma})^{-1}, \text{ where } n_i = \dim A_i.$$

(d) $A = \mathbb{Z}[T_1, \ldots, T_n]$. It follows from the calculations in (a) that in this case

$$\zeta(s) = \prod_p \frac{1}{1 - p^{n-s}} = \zeta(s - n)$$

is the usual ζ-function with shifted argument whose Euler product, as is well known, converges absolutely for Re $(s - n) > 1$, i.e., Re $s > n + 1 = \dim A$.

(e) Let A be a ring without zero divisors containing $\mathbb{Z}$. If we can find a subring $\mathbb{Z}[T_1, \ldots, T_n]$ of A, over which A is integer, the arguments in (b) will yield the result. Regrettably, this is not always possible; we can, however, remedy the situation for the price of localization modulo a finite number of primes.

More precisely, let us apply Noether's normalization theorem, see Theorem 1.5.3f, to $A' = \mathbb{Q} \otimes_{\mathbb{Z}} A$ and find a subring $\mathbb{Q}[T_1, \ldots, T_n]$ in A' over which A is integer. Multiplying, if necessary, the T_i by integers, we may assume that $T_i \in A$. Any element of A over $\mathbb{Z}[T_1, \ldots, T_n]$ satisfies an equation whose leading coefficient is an integer. Consider the set of prime divisors of all such leading coefficients for a finite family of generators of A over $\mathbb{Z}[T_1, \ldots, T_n]$ and denote by S the multiplicative set generated by this set. Then A_S is integral over $\mathbb{Z}_S[T_1, \ldots, T_n]$, and

$$\zeta_A(s) = \prod_{p \in S} \zeta_{A/(p)}(s) \prod_{p \notin S} \zeta_{A/(p)}(s). \tag{1.22}$$

The set $\{p \mid p \in S\}$ is finite and, by the above, $\zeta_{A/(p)}(s)$ converges uniformly for Re $s > \dim A/(p) \geq \dim A - 1$.

For $p \notin S$, we have $\zeta_{A/(p)}(s) = \zeta_{A_S/(p)}(s)$, and the constant α for the pair of rings $\mathbb{Z}_S[T_1, \ldots, T_n]/(p) \subset A_S/(p)$ introduced in (b) can be chosen to be independent of p. Indeed, the class modulo p of the fixed family of integer generators of A_S over $\mathbb{Z}_S[T_1, \ldots, T_n]$ gives a family of generators of $A_S/(p)$ for all $p \notin S$. Therefore, for $\sigma = $ Re $s > \dim A/(p)$ the second (infinite) product in (1.22) is majorized by

$$\prod_{p \notin S} (1 - p^{n-\sigma})^{-\alpha},$$

and therefore converges uniformly for $\sigma > n + 1 = \dim A$.

(f) Finally, we trivially reduce the general case to the ones treated above by decomposing Spec A into irreducible components as in (c).   □

1.14.6 Exercises

(1) Express $n(p^a)$ in terms of $v(p^b)$.
(2) Calculate the number of irreducible polynomials of degree d in one indeterminate over the field of q elements.
(3) Calculate $\zeta_A(s)$, where $A = \mathbb{Z}[T_1, \ldots, T_n]/(F)$ and F is a quadratic form.
(4) Let A be a ring of finite type over $\mathbb{Z}$, and $P \subset \mathbb{N}$ the set of prime numbers; let

$$S = \{ p \in P \mid \text{there exists } x \in \operatorname{Spec} A, \text{ such that } \operatorname{Char} k(x) = p \}.$$

Prove that either S is finite, or $P \setminus S$ is finite. For an integral domain not of finite type over $\mathbb{Z}$, give an example when both S and $P \setminus S$ are infinite.

1.15   Affine Group Schemes

In this section, I give definitions and several of the most important examples of affine group schemes. This notion is not only important by itself, it also transparently shows the role and capabilities of the "categorical" and "structural" approaches.

We will give two definitions of a group structure on an object of a category and compare them for the category of schemes.

1.15.1   Group Structure on an Object of a Category

Let C be a category, $X \in \operatorname{Ob}\mathsf{C}$. We say that on X there is given a *group structure* if on all sets $P_X(Y) := \operatorname{Hom}_\mathsf{C}(Y, X)$, called the *sets of Y-points of X*, there are given (set-theoretical) group structures, and, for any morphism $Y_1 \longrightarrow Y_2$, the corresponding map of sets $P_X(Y_2) \longrightarrow P_X(Y_1)$, given by composition, is a group homomorphism.

An object X together with a group structure on it is called a *group in the category* C. Let X_1 and X_2 be groups in C; a morphism $X_1 \longrightarrow X_2$ in C is a *group morphism in* C if the maps $P_{X_1}(Y) \longrightarrow P_{X_2}(Y)$ are group homomorphisms for any Y.

A group in the category of affine schemes will be called an *affine group scheme*.[24]

Below is a list of the most important examples with their standard notation and names.

---

[24]But never an *affine group*: this term is reserved for a different notion.

1.15.1a Helpful Remark

Since $\mathsf{Aff\,Sch}^{\circ}$, the category dual to the category of affine schemes $\mathsf{Aff\,Sch}$, is equivalent to Rings, instead of studying contravariant functors on $\mathsf{Aff\,Sch}$ represented by an affine group scheme we may discuss the covariant functors on Rings, which are simpler to handle.

1.15.2 Examples

1.15.2a The Additive Group $\mathbb{G}_{\mathrm{a}} = \mathrm{Spec}\,\mathbb{Z}[T]$

Any morphism $\mathrm{Spec}\,A \longrightarrow \mathbb{G}_{\mathrm{a}}$ is uniquely determined by an element $t \in A$, the image of T, which may be chosen arbitrarily. The collection of groups with respect to addition $A = \mathbb{G}_{\mathrm{a}}(A)$ for the rings $A \in \mathsf{Rings}$ defines the group structure on $\mathbb{G}_{\mathrm{a}}$.

In other words, $\mathbb{G}_{\mathrm{a}}$ represents the functor $\mathsf{Aff\,Sch}^{\circ} \longrightarrow \mathsf{Gr}$ given by the correspondences $\mathrm{Spec}\,A \longmapsto A$ or, equivalently, the functor $\mathsf{Rings} \longrightarrow \mathsf{Gr}$ given by $A \longmapsto A^{+}$, the latter being the group A with respect to $+$.

1.15.2b The Multiplicative Group $\mathbb{G}_{\mathrm{m}} = \mathrm{Spec}\,\mathbb{Z}[T, T^{-1}]$

For any scheme $X = \mathrm{Spec}\,A$, any morphism $X \longrightarrow \mathbb{G}_{\mathrm{m}}$ is uniquely determined by an element $t \in A^{\times}$, the image of T under the homomorphism $\mathbb{Z}[T, T^{-1}] \longrightarrow A$; here $A^{\times}$ is the group (with multiplication as the group operation) of invertible elements of the ring A. Conversely, t corresponds to such a morphism if and only if $t \in A^{\times}$. Therefore

$$P_{\mathbb{G}_m}(\mathrm{Spec}\,A) = \mathbb{G}_{\mathrm{m}}(A) = A^{\times},$$

and on this set of A-points of $\mathbb{G}_{\mathrm{m}}$ the natural group structure (multiplication) is defined. Furthermore, any ring homomorphism $A \longrightarrow B$ clearly induces a group homomorphism $A^{\times} \longrightarrow B^{\times}$ which defines the group structure on $\mathbb{G}_{\mathrm{m}}$.

In other words, $\mathbb{G}_{\mathrm{m}}$ represents the functor $\mathsf{Aff\,Sch}^{\circ} \longrightarrow \mathsf{Gr}$ given by the correspondences $\mathrm{Spec}\,A \longmapsto A^{\times}$ or, equivalently, the functor $\mathsf{Rings} \longrightarrow \mathsf{Gr}$ given by $A \longmapsto A^{\times}$.

1.15.2c The General Linear Group

$$\mathrm{GL}(n) = \mathrm{Spec}\,\mathbb{Z}[T_{ij}, T]_{i,j=1}^{n}/(T\det(T_{ij}) - 1). \tag{1.23}$$

It represents the functor $\mathrm{Spec}\,A \longmapsto \mathrm{GL}(n; A)$. Obviously, $\mathrm{GL}(1) \simeq \mathbb{G}_{\mathrm{m}}$.

1.15.2d   The Galois Group Aut(K'/K)

Fix a K-algebra K' such that K' is a free K-module of finite rank. The group
Aut(K'/K) of automorphisms of the algebra K' over K is the main object of study,
e.g., in Galois theory (where only the case of fields K, K' is considered). This group
may turn to be trivial if the extension is non-normal or non-separable, and so on.

The functorial point of view suggests to consider all possible *changes of base K*,
i.e., for a variable K-algebra B, consider the group of automorphisms

$$\text{Aut}(B'/B) := \text{Aut}_B(B'), \quad \text{where } B' = B \otimes_K K'.$$

We will prove simultaneously that (1) the map $B \longmapsto \text{Aut}(B'/B)$ is a functor and (2)
this functor is representable.

Select a free basis $e_1, \ldots, e_n$ of K' over K. In this basis, the multiplication law in
K' is given by the formula

$$e_i e_j = \sum_{1 \le k \le n} c_{ij}^k e_k.$$

Denote $e_i' := 1 \otimes_K e_i$; then $B' = \bigoplus_{1 \le i \le n} B e_i'$, and any endomorphism t of the
B-module B' is given by a matrix (t_{ij}), where $t_{ij} \in B$ and $1 \le i, j \le n$. The condition
that this matrix determines an endomorphism of an algebra can be expressed as the
relations

$$t(e_i')t(e_j') = \sum_{1 \le k \le n} c_{ij}^k t(e_k'). \tag{1.24}$$

Equating the coefficients of e_k' in (1.24) in terms of indeterminates T_{ij} we obtain
a system of algebraic relations for T_{ij} with coefficients in K, both necessary and
sufficient for (t_{ij}) to define an endomorphism of B'/B.

To obtain *automorphisms*, we introduce an additional indeterminate T and the
additional relation, cf. Example of the general linear group GL(n), Eq. (1.23), which
ensures that $\det(T_{ij})$ does not vanish:

$$T \det(T_{ij}) - 1 = 0.$$

The quotient $K[T, T_{ij}]_{i,j=1}^n / (T \det(T_{ij})_{i,j=1}^n - 1)$ is a K-algebra representing the
functor

$$B \longmapsto \text{Aut}(B'/B).$$

This K-algebra replaces the notion of the *Galois group* of the extension K'/K; it
generalizes the notion of the group ring of the Galois group.

1.15.2e Particular Case: Quadratic Extension of the Field

Consider the simplest particular case:

$$K' = K(\sqrt{a}), \text{ where } a \in K^{\times} \setminus (K^{\times})^2.$$

We may set $e_1 = 1$, $e_2 = \sqrt{a}$; the multiplication table reduces to $e_2^2 = a$.

Let $t(\sqrt{a}) = T_1 + T_2\sqrt{a}$ (obviously, $t(1) = 1$). Since $t(\sqrt{a})^2 = a$, we obtain the equations relating T_1, T_2 and the additional variable T:

$$\begin{cases} T_1^2 + a\,T_2^2 & = a, \\ 2T_1\,T_2 & = 0, \\ TT_2 - 1 & = 0. \end{cases}$$

We consider separately two cases.

Case 1 Char $K \neq 2$. Hence, 2 is invertible in any K-algebra. The functor of automorphisms is represented by the K-algebra

$$K[T, T_1, T_2]/(T_1^2 + a\,T_2^2 - a,\ T_1T_2,\ TT_2 - 1).$$

If B has no zero divisors, then the B-points of this K-algebra have a simple structure: because T_2 must not vanish, it follows that T_1 vanishes, implying that the only possible values of T_2 in the quotient ring are ± 1. Like the usual ("ordinary") Galois group, this group is isomorphic to $\mathbb{Z}/2\mathbb{Z}$; the automorphisms simply change the sign of $\sqrt{a}$.

The following case illustrates that when B does have zero divisors the group of B-points of Aut $K^{\times}/K$ can be considerably larger.

Case 2 Char $K = 2$. The functor of automorphisms is represented by the K-algebra

$$K[T, T_1, T_2]/(T_1^2 + aT_2^2 - a, TT_2 - 1).$$

In other words, the B-points of the automorphism group are all B-points of the circle $T_1^2 + aT_2^2 - a = 0$ at which T_2 is invertible!

Let us examine this closer. Let B be a field and let (t_1, t_2) be a B-point of the circle at which T_2 is invertible. Then either $t_2 = 1$, $t_1 = 0$, and we obtain the identity automorphism, or $a = \left(\dfrac{t_1}{t_2 + 1}\right)^2$. Therefore, there are nontrivial B-points only if $\sqrt{a} \in B$, in which case the equation of the circle turns into the square of a linear equation

$$(T_1 + \sqrt{a}\,T_2 + \sqrt{a})^2 = 0.$$

We have a punctured line (the line without point $T_2 = 0$) of automorphisms!

Obviously, $\text{Aut}(B'/B)$ is isomorphic in this case to $B^\times$, the multiplicative group of B (under the composition of automorphisms the coefficients of $\sqrt{a}$ are multiplied). So, the non-separable extensions have even more, in a certain sense, automorphisms than separable ones.

The reason for this phenomenon is the appearance of nilpotents in the algebra $B \otimes_K K'$. Indeed, $K(\sqrt{a}) \subset B$, so $K(\sqrt{a}) \otimes_K K(\sqrt{a}) \subset B \otimes_K K'$; on the other hand, this product is isomorphic to

$$K(\sqrt{a})[x]/(x^2 - a) \simeq K(\sqrt{a})[y]/(y^2).$$

The automorphisms just multiply y by invertible elements.

One can similarly investigate arbitrary finite nonseparable extensions and construct for them an analog of the Galois theory.

1.15.2f   The Group μ_n of nth Roots of Unity

Set

$$\mu_n = \text{Spec } \mathbb{Z}[T]/(T^n - 1) = \text{Spec } \mathbb{Z}[T, T^{-1}]/(T^n - 1).$$

This group represents the functor $\text{Spec } A \longmapsto \{t \in A^\times \mid t^n = 1\}$.

Let X be a closed affine group scheme and Y a closed subscheme of X such that $P_Y(Z) \subset P_X(Z)$ is a subgroup for any Z. We call Y with the induced group structure a *closed subgroup* of X.

Clearly, μ_n is a closed subgroup of $\mathbb{G}_m$. Explicitly, the homomorphism $T \longmapsto T^n$ determines the group scheme homomorphism $\mathbb{G}_m \longrightarrow \mathbb{G}_m$ of "raising to the power n" and μ_n represents the kernel of this homomorphism.

1.15.2g   The Scheme of a Finite Group G

Let G be an ordinary (set-theoretical) finite group. Set $A = \mathbb{Z}^{(G)} := \prod_{g \in G} \mathbb{Z}$. In other words, A is a free module $\bigoplus_{g \in G} \mathbb{Z}^{(g)}$, i.e., $|G|$ copies of $\mathbb{Z}$, with the multiplication table

$$e_g e_h = \begin{cases} 0 = (0, \ldots, 0), & \text{if } h \neq g, \\ e_g, & \text{if } h = g. \end{cases}$$

The space $X = \text{Spec } A$ is disconnected; each of its components is isomorphic to $\text{Spec } \mathbb{Z}$ and these components are indexed by the elements of G. For any ring B whose spectrum is connected, the set of morphisms $\text{Spec } B \longrightarrow \text{Spec } A$ is, therefore, in a natural one-to-one correspondence with the elements of G.

If Spec B is disconnected, then a morphism Spec $B \longrightarrow$ Spec A is determined by the set of its restrictions to the connected components of Spec B. Let Conn B be the set of these components; then, clearly, the point functor is given by

$$P_X(\text{Spec } B) \xrightarrow{\simeq} (G)^{\text{Conn } B} := \text{Hom}(G, \text{Conn } B),$$

and therefore X is endowed with a natural group structure called the *scheme of the finite group G*.

1.15.2h The Relative Case

Let $S = \text{Spec } K$. A group object in the category Aff Sch$_S$ of affine schemes over S is called an *affine S-group* (or an *affine K-group*). Setting $\mathbb{G}_{m/K} = \mathbb{G}_m \times S$ and $\mu_{n/K} = \mu_n \times S$, and so on, we obtain a series of groups over an arbitrary scheme S (or the corresponding ring K). Each of these groups represents "the same" functor as the corresponding absolute group, but restricted to the category of K-algebras.

1.15.2i Linear Algebraic Groups

Let K be a field. Any closed subgroup of $\text{GL}_n(A)_{/K}$ is called a *linear algebraic group over K*.

In other words, a linear algebraic group is defined by a system of equations

$$F_k(T_{ij}) = 0, \quad \text{for } i,j = 1, \dots, n \text{ and } k \in I, \tag{1.25}$$

such that if (t'_{ij}) and (t''_{ij}) are two solutions of the system (1.25) in a K-algebra A such that the corresponding matrices are invertible, then the matrix $(t'_{ij})(t''_{ij})^{-1}$ is also a solution of (1.25)

The place of linear algebraic groups in the general theory is elucidated by the following fundamental theorem (cf. [OV, Theorem 8 of Chapter 3], the same is proved for algebraic groups instead of affine group schemes).

1.15.2i.i Theorem *Let X be an affine group scheme of finite type over K. Then X is isomorphic to a linear algebraic group.*

Recall that Hermann Weyl referred to the general linear group as "Her All-embracing Majesty".

1.15.3 Theorem (Cartier) *Let X be the scheme of a linear algebraic group scheme over a field of characteristic zero. Then X is reduced, i.e., $X = X_{red}$, its ring has no nilpotents.*

For a proof of this theorem, see, e.g., [M1].   □

The statement of the theorem is false if[25] Char $K = p > 0$ as demonstrated by the following example..

1.15.3i Example Set

$$\mu_{p\,/K} = \operatorname{Spec} K[T]/(T^{p-1}) = \operatorname{Spec} K[T]/((T-1)^p). \qquad (1.26)$$

Obviously, $K[T]/((T-1)^p)$ is a local Artinian algebra of length p, and its spectrum should be considered as a "point of multiplicity p". This agrees nicely with our intuition: all roots of unity of degree p are glued together and turn into one root of multiplicity p.

More generally, set

$$\mu_{p^n\,/K} = \operatorname{Spec} K[T]/((T-1)^{p^n}).$$

We see that the length of the nilradical may be arbitrarily large.

1.15.4   The Set-theoretical Definition of the Group Structure

Let the category C contain a final[26] object E and products (e.g., $X \times X$ and $X \times X \times X$).

The usual set-theoretical definition of the group structure on a set X given above is, clearly, equivalent to the existence of three morphisms

$$m : X \times X \longrightarrow X \quad (\text{multiplication}, (x, y) \longmapsto xy)$$

$$i : \quad X \quad \longrightarrow X \quad (\text{inversion}, x \longmapsto x^{-1})$$

$$u : \quad E \quad \longrightarrow X \quad (\text{the embedding of } E, \text{ i.e, } E \longmapsto 1)$$

that satisfy the axioms of associativity, left inversion and left unit, respectively, expressed as commutativity of the following diagrams, where "$\bullet$" in (1.28) and (1.29) is the morphism of contraction to the point (E):

$$
\begin{array}{ccc}
X \times X \times X & \xrightarrow{\;(m,\,\mathrm{id}_X)\;} & X \times X \\
{\scriptstyle (\mathrm{id}_X,m)}\downarrow & & \downarrow{\scriptstyle m} \\
X \times X & \xrightarrow{\quad m \quad} & X
\end{array}
\qquad (1.27)
$$

---

[25]It is also false for group **super**schemes in any characteristic.

[26]An object E is said to be *final* if $\mathrm{card}(\mathrm{Hom}(X, E)) = 1$ for any $X \in \mathrm{Ob}\,C$.

$$X \times X \xrightarrow{(i,\ \mathrm{id}_X)} X \times X$$

$$\begin{array}{ccc} X \times X & \xrightarrow{(i,\ \mathrm{id}_X)} & X \times X \\ \uparrow{\scriptstyle \delta} & & \downarrow{\scriptstyle m} \\ X \xrightarrow{\ \bullet\ } E \xrightarrow{\ u\ } X \end{array} \tag{1.28}$$

$$\begin{array}{ccccc} X \times X & \xrightarrow{(\bullet,\ \mathrm{id}_X)} & E \times X & \xrightarrow{(u,\ \mathrm{id}_X)} & X \times X \\ \uparrow{\scriptstyle \delta} & & & & \downarrow{\scriptstyle m} \\ X & \xrightarrow{\ \mathrm{id}_X\ } & X & \xrightarrow{\ \mathrm{id}_X\ } & X \end{array} \tag{1.29}$$

1.15.4a Exercise Formulate the commutativity axiom for the group.

In the category Sets, the axioms (1.27)–(1.29) turn into the usual definition of a group, though in an somewhat non-conventional form. Let X be a group with a unit 1; let $x, y, z \in X$; then, in the standard notation, we have

$$m(x, y) = xy, \quad i(x) = x^{-1}, \quad u(E) = 1,$$

and the conventional axioms of associativity, left inverse and left unit have the respective forms

$$(xy)z = x(yz), \quad x^{-1}x = 1, \quad 1x = x.$$

1.15.5 Equivalence of the Two Definitions of Group Structure

Let a group structure in the set-theoretical sense be given on $X \in \mathrm{Ob}\,\mathsf{C}$. Then, for every $Y \in \mathrm{Ob}\,\mathsf{C}$, the morphisms m, i, u induce a group structure on the set $P_X(Y)$ of Y-points.

1.15.5a Exercise Verify the compatibility of these structures with the maps $P_X(Y_1) \longrightarrow P_X(Y_2)$ induced by the maps $Y_1 \longrightarrow Y_2$.

Conversely, let a group structure in the sense of the first definition be given on $X \in \mathrm{Ob}\,\mathsf{C}$. How to recover the morphisms m, i, u? We do it in three steps:

(a) The group $P_X(X \times X)$ contains the projections $\pi_1, \pi_2 : X \times X \longrightarrow X$.
  Set $m := \pi_1 \cdot \pi_2$ (the product $\cdot$ in the sense of the group law).
(b) The group $P_X(X)$ contains the element id_X. Denote its inverse (in the sense of the group law) by i.
(c) The group $P_X(E)$ has a unit element. Denote it by $u : E \longrightarrow X$.

1.15.5a.i Exercise (1) Prove that m, i, u satisfy the axioms of the second definition.
(2) Verify that the constructed maps of sets

$$
\left\{
\begin{array}{c}
\text{group structures on } X \\
\text{in the sense of} \\
\text{the first definition}
\end{array}
\right\}
\longleftrightarrow
\left\{
\begin{array}{c}
\text{group structures on } X \\
\text{in the sense of} \\
\text{the second definition}
\end{array}
\right\}
\qquad (1.30)
$$

are mutually inverse.

1.15.6 Description of the Group Structure on an Affine Group Scheme $X = \operatorname{Spec} A$ in Terms of A

We will consider directly the relative case, i.e., assume A to be a K-algebra.

The notion of a group G is usually formulated in terms of states, i.e., points of G. In several questions, however, for example, to quantize it, we need a reformulation in terms of observables, i.e., functions on G. Since any map of sets $\varphi: X \longrightarrow Y$ induces a homomorphism of the algebras of functions $\varphi^*: F(Y) \longrightarrow F(X)$, we may dualize the axioms of Sect. 1.15.4 and obtain the following definition.

A *bialgebra* structure on a K-algebra A is given by three K-algebra homomorphisms of a *coalgebra*:

$$
\begin{aligned}
m^* &: A \longrightarrow A \otimes_K A \quad \text{(comultiplication)} \\
i^* &: A \longrightarrow A \qquad\quad \text{(coinversion)} \\
u^* &: A \longrightarrow K \qquad\quad \text{(counit)}
\end{aligned}
$$

which satisfy the axioms of coassociativity, left coinversion, and left counit, respectively, expressed as commutativity of the following diagrams:

(the left vertical arrow is the multiplication $\mu : a \otimes b \longmapsto ab$ in A, the left horizontal arrow is given by $1 \longmapsto 1$).

$$
\begin{array}{ccccc}
A \otimes A & \longleftarrow & A & \xleftarrow{\;u^* \otimes \mathrm{id}_A\;} & A \otimes A \\
\Big\downarrow{\scriptstyle \mu} & & & & \Big\uparrow{\scriptstyle m^*} \\
A & \xleftarrow{\qquad \mathrm{id}_A \qquad} & & & A
\end{array}
$$

(the left top arrow is $a \longmapsto 1 \otimes a$).

It goes without saying that this definition is dual to that from Sect. 1.15.4, and therefore the group structures on the K-scheme $\mathrm{Spec}\,A$ are in a one-to-one correspondence with the co-algebra structures on the K-algebra A.

1.15.6a Example   The homomorphisms m^*, i^*, u^* for the additive group scheme $\mathbb{G}_{\mathrm{a}} = \mathrm{Spec}\,\mathbb{Z}[T]$ are:

$$
m^*(T) = T \otimes 1 + 1 \otimes T, \quad i^*(T) = -T, \quad u^*(T) = 0. \tag{1.31}
$$

1.15.6b Exercise   Write explicitly the homomorphisms m^*, i^*, u^* for the Examples 1.15.2.

1.16   Appendix: The Language of Categories. Representing Functors

1.16.1   General Remark

The language of categories is an embodiment of a "sociological" approach to mathematical objects: a group, a manifold or a space are considered not as a set with an intrinsic structure but as a member of a community of similar objects.

The "structural" and "categorical" descriptions of an object via the functor it represents are complementary. The role of the second description increases nowadays in various branches of mathematics (especially in algebraic geometry)[27] although its richness of content was first demonstrated in topology, by $K[\Pi, n]$ spaces.

---

[27]To say nothing of supersymmetric theories of theoretical physics, where the language of representable functors is a part of the working language (albeit often used sub- or un-consciously).

The proposed summary of definitions and examples purports to be an abridged phraseological dictionary of the language of categories (ordered logically rather than alphabetically).[28]

1.16.2 Definition of Categories

A *category* C is a collection of the following data:

(a) A set Ob C whose elements are called *objects of* C. (Instead of $X \in$ Ob C we often write briefly $X \in$ C.)

(b) For every ordered pair $X, Y \in$ C, there is given a (perhaps empty) set $\mathrm{Hom}_C(X, Y)$ (or briefly $\mathrm{Hom}(X, Y)$) whose elements are called *morphisms* or *arrows* from X into Y. Notation: $\varphi \in \mathrm{Hom}(X, Y)$ is written as $\varphi: X \longrightarrow Y$ and $\mathrm{Mor}\, C = \coprod_{X,Y \in C} \mathrm{Hom}(X, Y)$ is the collection of morphisms.

(c) For every ordered triple $X, Y, Z \in$ C, there is given a map

$$\mathrm{Hom}(X, Y) \times \mathrm{Hom}(Y, Z) \longrightarrow \mathrm{Hom}(X, Z)$$

assigning to morphisms $\varphi: X \longrightarrow Y$ and $\psi: Y \to Z$ a morphism $\psi \cdot \varphi$ (or just $\psi\varphi$) called their *composition*.

The data (a)–(c) must satisfy the following two axioms:

Associativity: $(\chi\psi)\varphi = \chi(\psi\varphi)$ for any $\varphi: X \longrightarrow Y$, $\psi: Y \longrightarrow Z$, and $\chi: Z \to V$.

Identity morphisms: for every $X \in$ C, there exists a morphism $\mathrm{id}_X: X \longrightarrow X$ such that $\mathrm{id}_X \cdot \varphi = \varphi$, $\psi \cdot \mathrm{id}_X = \psi$ whenever the compositions are defined. (Clearly, id_X is uniquely defined.)

A morphism $\varphi: X \longrightarrow Y$ is called an *isomorphism* if there exists a morphism $\psi: Y \longrightarrow X$ such that $\psi \cdot \varphi = \mathrm{id}_X$, $\varphi \cdot \psi = \mathrm{id}_Y$.

Given two categories C and D such that Ob C $\subset$ Ob D and

$$\mathrm{Hom}_C(X, Y) = \mathrm{Hom}_D(X, Y) \quad \text{for any } X, Y \in \text{Ob}\, C$$

and, moreover, such that the compositions of morphisms in C and in D coincide, one calls C a *full sub-category* of D.

---

[28] An excellent textbook is [McL]. New trends are reflected in [GM], see also http://www.math. jussieu.fr/~keller/publ/; see also Appendix in Molotkov's paper [Mo], where certain interesting subtleties, usually overlooked, are pointed at and clarified.

A category C is said to be *small* if $Ob\,C$ and the collections $Hom_C(X, Y)$ are sets. A category C is said to be *big* if $Mor\,C$ is a proper class[29] and, for any objects $X, Y \in C$, the class of morphisms $Hom_C(X, Y)$ is a set.

Sometimes, e.g., if we wish to consider a category of functors on a category C, such a category is impossible to define if $Ob\,C$ is a class; on the other hand, in the framework of small categories we cannot consider, say, the category of "all" sets, which is highly inconvenient.

P. Gabriel suggested a way out of this predicament: he introduced a *universum*, a large set of sets stable under all operations needed; then only categories belonging to this universum are considered. For a list of axioms a universum should satisfy see, e.g., Gabriel's thesis [Gab]. We will also assume the existence of a universum.

However, at the present state of the foundations of mathematics and of the problem of its consistency the whole problem seems to me a rather academic one. My position is close to that of an experimental physicist inclined neither to fetishize, nor to destroy his instruments as long they produce results.

Nicolas Bourbaki's opinion [Bb2] on this occasion reflects once again his Gaullean common sense and tolerance:

Mathematicians seem to agree to conclude also that there is only just a superficial concordance between our "intuitive" conceptions of the notion of set or whole number, and the formalisms which are supposed to account for them; the disagreement begins when it is a question of choosing between the one and the other.

1.16.3 Examples

For convenience we have grouped them as follows:

1.16.4 The First Group of Examples

The objects in this group are sets endowed with a structure, the morphisms are the maps of these sets preserving this structure. (A purist is referred for a definition of a *structure* to, e.g., [EM].)

---

[29]By definition, a *class* X is said to be a *set* if there exist Y such that $X \in Y$. Classes appear in BG (Bernays-Goedel (or Morse)) set theory; in the original ZF (Zermelo-Frenkel) set theory only sets exist. But, if one adds to ZF an axiom of existence, for any set X, of a universum U containing X (as Grothendieck did in [EGA, Gab]), then BG theory is modeled inside ZF with this additional axiom: The sets of BG are modeled by sets X of ZF+Bourbaki belonging to some universum U big enough to contain the set of natural numbers, whereas classes of BG are subsets of U.

But what is a universum? It is any set closed with respect to the ordinary set-theoretic operations: unions of (a family of) sets, intersections, power set, etc. (for a precise definition, see [EGA, Gab]). See also Appendix in [Mo].

- Sets, or Ens for French-speakers, the category of sets and their maps;
- Top, the category of topological spaces and their continuous maps;
- Gr, the category of groups and their homomorphisms;
- Gr$_f$, the subcategory of Gr whose objects are the finite (as the subscript indicates) groups;
- Ab, the subcategory of Gr of Abelian groups;
- Aff Sch, the category of affine schemes and their morphisms;
- Rings, the category of (commutative) rings (with unit) and their unit preserving homomorphisms;
- A-Algs, the category of algebras over an algebra A;
- A-Mods, the category of (left) A-modules over a given algebra A;
- Man, the category of manifolds and their morphisms.

1.16.5   The Second Group of Examples

The objects in this group are still structured sets, but the morphisms are no longer structure-preserving maps of these sets. (No fixed name for some of these categories.)

- The main category of homotopic topology: its objects are topological spaces, the morphisms are the homotopy classes of continuous maps (see, e.g., [FFG]).
- *Additive relations* (see, e.g., [GM]; no fixed name for this category): its objects are Abelian groups. A morphism $f: X \longrightarrow Y$ is any subgroup of $X \times Y$, and the composition of $\varphi: X \longrightarrow Y$ and $\psi: Y \longrightarrow Z$ is given by the relation

$$\psi\varphi = \{(x, z) \in X \times Z \mid \text{ there exists } y \in Y \text{ such that}$$

$$(x, y) \in \varphi, (y, z) \in \psi\}.$$

1.16.6   The Third Group of Examples

The examples in this group are constructed from some classical structures that one finds *sometimes* convenient to view as categories.

- For a (partially) ordered set I, the category $\mathsf{C}(I)$ is given by $\mathrm{Ob}\,\mathsf{C}(I) = I$, and

$$\mathrm{Hom}_{\mathsf{C}(I)}(X, Y) = \begin{cases} \text{one element, if } X \leq Y, \\ \emptyset, \qquad\qquad \text{otherwise.} \end{cases}$$

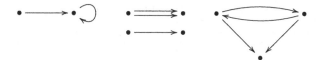

Fig. 1.13

The main example: I is the set of indices of an *inductive system* of groups.[30]

- Let X be a topological space. Denote by Top_X the category whose objects are the open sets of X and whose morphisms are their natural embeddings. This trivial reformulation conceives the germ of an astoundingly deep generalization of the notion of a topological space, the *Grothendieck topologies* or *toposes*, see [J].[31]

- *Category associated with a diagram scheme aka quiver.*

  A *diagram scheme* is (according to Grothendieck) a triple consisting of two sets, V (vertices) and A (arrows), and the map $e: A \longrightarrow V \times V$ that to every arrow of A assigns its endpoints, i.e., an ordered pair of vertices: the source and the target of the arrow (Fig. 1.13).

  Let (V, A, e) be a graph. Define the category $\mathsf{D} = \mathsf{D}(V, A, e)$ setting $\mathrm{Ob}\,\mathsf{D} = V$ and $\mathrm{Hom}_\mathsf{D}(X, Y) = \{$paths from X to Y along the arrows$\}$.

  More precisely, if $X \neq Y$, then any element of $\mathrm{Hom}_\mathsf{D}(X, Y)$ is a sequence $f_1, \ldots, f_k \in A$ such that the source of f_1 is X, the target of f_i is the source of f_{i+1}, and the target of f_k is Y. If $X = Y$, then $\mathrm{Hom}_\mathsf{D}(X, X)$ must contain the identity arrow. The compositions of morphisms is defined in an obvious way as the composition of paths.

- Define the category $\mathsf{D}_C = \mathsf{D}_C(V, A, e)$ by setting $\mathrm{Ob}\,\mathsf{D}_C = V$ and

$$\mathrm{Hom}_{\mathsf{D}_C}(X, Y) = \begin{cases} \text{one element,} & \text{if } \mathrm{Hom}_\mathsf{D}(X, Y) \neq \emptyset, \\ \emptyset, & \text{otherwise.} \end{cases} \tag{1.32}$$

---

[30]Let $(I, \leq)$ be a directed poset (not all authors require I to be directed, i.e., to be nonempty and equipped with a reflexive and transitive binary relation $\leq$, with the additional property that every pair of elements has an upper bound). Let $(G_i)_{i \in I}$ be a family of groups and assume that

(1) one has a family of homomorphisms $f_{ij} : G_i \longrightarrow G_j$ for all $i \leq j$ such that:
(2) f_{ii} is the unit element in G_i,
(3) $f_{ik} = f_{ij} \circ f_{jk}$ for all $i \leq j \leq k$.

Then the set of pairs $(G_i, f_{ij})_{i,j \in I}$ is called an *inductive system* (or *direct system*) of groups and morphisms over I.

If in assumption (1) above we have a family of homomorphisms $f_{ij} : G_j \longrightarrow G_i$ for all $i \leq j$ (note the order) with the same properties, then the set of pairs $(G_i, f_{ij})_{i,j \in I}$ is called an *inverse system* (or *projective system*) of groups and morphisms over I.

[31]A. Rosenberg used it to construct spectra of noncommutative rings in his "noncommutative" books [R1], [R2] and MPIM-Bonn preprints with M. Kontsevich.

Intuitively, D_C is the category corresponding to a commutative diagram: all paths from X to Y define the same morphism.

1.16.4 Examples of Constructions of Categories

There are several useful formal constructions of new categories. I will describe here only three of them.

1.16.4a The Dual Category

Given a category C, define its *dual* C° by setting $\mathrm{Ob}\,C^\circ$ to be a copy of $\mathrm{Ob}\,C$ and $\mathrm{Hom}_{C^\circ}(X^\circ, Y^\circ)$ to be in one-to-one correspondence with $\mathrm{Hom}_C(Y, X)$, where $X^\circ \in \mathrm{Ob}\,C^\circ$ denotes the object corresponding to $X \in \mathrm{Ob}\,C$, so that if a morphism $\varphi^\circ \colon X^\circ \longrightarrow Y^\circ$ corresponds to a morphism $\varphi \colon Y \longrightarrow X$, then $\psi^\circ \varphi^\circ = (\varphi\psi)^\circ$ and $\mathrm{id}_{X^\circ} = (\mathrm{id}_X)^\circ$.

Informally speaking, C° is obtained from C by taking the same objects, but inverting the arrows.

This construction is interesting in two opposite cases. If C° highly resembles C, e.g., is equivalent (the definition of equivalence of categories will be given a little later) to C (as is the case of, say $\mathsf{Ab_f}$, the category of finite Abelian groups), this situation provides us with a stage where various duality laws perform.

Conversely, if C° is quite unlike C, then the category C° might have new and nice properties compared to C; e.g., for $C = \mathsf{Rings}$, the dual category $C^\circ = \mathsf{Aff\,Sch}$ has "geometric" qualities enabling us to glue global objects from local ones—an operation appallingly unnatural, and even impossible to imagine, inside Rings.

1.16.4b The Category of Objects over a Given Base

Given a category C and an object S in it, set $\mathrm{Ob}\,C_S = \{\varphi \in \mathrm{Hom}_C(X, S)\}$. For any $\varphi \colon X \longrightarrow S$ and $\psi \colon Y \longrightarrow S$, define:

$$\mathrm{Hom}_{C_S}(\varphi, \psi) = \{\chi \in \mathrm{Hom}_C(X, Y) \mid \varphi = \psi\chi\},$$

i.e., $\mathrm{Hom}_{C_S}(\varphi, \psi)$ is the set of commutative diagrams

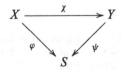

The composition of morphisms in C_S is induced by the composition in C.

1.16.4c Examples

(1) $Rings_R = R$-Algs, the category of R-algebras, where R is a fixed ring;
(2) $Vebun_M$, the category of vector bundles over a fixed base M; this is a subcategory of the category Bun_M whose fibers are arbitrary manifolds;
(3) the category $(C_S)^\circ$ that deals with morphisms $S \longrightarrow X$ for a fixed $S \in C$.

1.16.4d   The Product of Categories

Given a family of categories C_i, where $i \in I$, define their product $\prod_{i \in I} C_i$ by setting

$$Ob \prod C_i = \prod Ob\, C_i,$$
$$Hom_{\prod C_i}\left(\prod X_i, \prod Y_i\right) = \prod Hom_{C_i}(X_i, Y_i)$$

with coordinate-wise composition.

1.16.5   Functors

A *covariant functor*, or just a *functor*, $F: C \longrightarrow D$, from the category C to the category D is a collection of maps $(F, F_{X,Y})$, usually abbreviated to F, where $F: Ob\, C \longrightarrow Ob\, D$ and

$$F_{X,Y}: Hom_C(X, Y) \longrightarrow Hom_D(F(X), F(Y)) \quad \text{for any } X, Y \in C,$$

such that $F_{X,Z}(\varphi\psi) = F_{Y,Z}(\varphi)F_{X,Y}(\psi)$ for any $\varphi, \psi \in Hom\, C$ provided $\varphi\psi$ is defined.

A functor from C° to D is called a *contravariant functor* from C to D. A functor $F: C_1 \times C_2 \longrightarrow D$ is called a *bifunctor*, and so on.

Given categories C, D, E and functors $F: C \longrightarrow D$ and $G: D \longrightarrow E$, define $GF: C \to E$ composing the constituents of F and G in the usual set-theoretical sense.

The most important examples of functors are just "natural constructions": (co)homology and homotopy are functors $Top \longrightarrow Ab$; the characters of finite groups constitute a functor $Gr_f \longrightarrow Rings$. These examples are too meaningful to discuss them in passing.

1.16.6   Examples of Presheaves

A *presheaf of sets* (groups, rings, algebras, superalgebras, R-modules, and so on) is
a contravariant functor from the category Top_X of open sets of a topological space
X to Sets (or Gr, Rings, Algs, Salgs, $R\text{-}\mathsf{Mods}$, and so on, respectively).

Let (V, A, e) be a diagram scheme, D and D_C the associated categories. A functor
from D to a category E is called a *diagram of objects from* E (of type (V, A, e)). A
functor from D_C to E is a *commutative* diagram of objects.

For an ordered set I considered as a category, a functor from I to a category C is
a family of objects in C indexed by the elements of I and connected with morphisms
so that these objects constitute either a projective or an inductive system in C.

Given two functors $F, G\colon \mathsf{C} \longrightarrow \mathsf{D}$, a *natural transformation* or a *morphism of
functors* $f\colon F \longrightarrow G$ is a set of morphisms $f(X)\colon F(X) \to G(X)$ (one for each $X \in \mathsf{C}$)
such that for any $\varphi \in \mathrm{Hom}_C(X, Y)$ the following diagram commutes:

$$
\begin{array}{ccc}
F(X) & \xrightarrow{\ f(X)\ } & G(X) \\
{\scriptstyle F(\varphi)}\downarrow & & \downarrow{\scriptstyle G(\varphi)} \\
F(Y) & \xrightarrow{\ f(Y)\ } & G(Y).
\end{array}
$$

The *composition of natural transformations* is naturally defined.

A natural transformation f is called a *functor isomorphism* if $f(\varphi) \in \mathrm{Mor}\,\mathsf{D}$ are
isomorphisms for all $X \in \mathsf{C}$. The functors from C to D are objects of a category
denoted by $\mathsf{Funct}(\mathsf{C}, \mathsf{D})$.

A functor $F\colon \mathsf{C} \longrightarrow \mathsf{D}$ is called an *equivalence of categories* if there exists
a functor $G\colon \mathsf{D} \longrightarrow \mathsf{C}$ such that $GF \cong \mathrm{id}_C$, $FG \cong \mathrm{id}_D$; in this case C is said to
be *equivalent* to D.

1.16.6a Examples

(1) $(\mathsf{Ab_f})^\circ \cong \mathsf{Ab_f}$ so that $G \longleftrightarrow \mathcal{X}(G)$, the character group of G;
(2) Rings° is equivalent to the category $\mathsf{Aff\ Sch}$ of affine schemes.

1.16.7   The Category $C^* := Funct(C^\circ, Sets)$

If our universum is not too large, there exists a category whose objects are categories
and whose morphisms are functors between them. The main example: the category
$\mathsf{C}^* = \mathsf{Funct}(\mathsf{C}^\circ, \mathsf{Sets})$ of functors from C° to Sets.

1.16.8 Representable Functors

Fix any $X \in \mathsf{C}$.

(1) Denote by $P_X \in \mathsf{C}^*$ (here P stands for point; usually this functor is denoted by h_X, where h stands for "homomorphism") the functor given by

$$P_X(Y^\circ) = \mathrm{Hom}_{\mathsf{C}}(Y, X) \quad \text{for any } Y^\circ \in \mathsf{C}^\circ;$$

to any morphism $\varphi^\circ: Y_2^\circ \longrightarrow Y_1^\circ$ the functor P_X assigns the map of sets $P_X(Y_2^\circ) \longrightarrow P_X(Y_1^\circ)$ which sends $\psi: Y_2 \longrightarrow X$ into $\varphi\psi: Y_1 \longrightarrow Y_2 \longrightarrow X$.

To any $\varphi \in \mathrm{Hom}_{\mathsf{C}}(X_1, X_2)$, there corresponds a functor morphism $P_\varphi: P_{X_1} \longrightarrow P_{X_2}$ which to every $Y \in C$ assigns

$$P_\varphi(Y^\circ): P_{X_1}(Y^\circ) \longrightarrow P_{X_2}(Y^\circ)$$

and sends every morphism $\psi \in \mathrm{Hom}_{\mathsf{C}}(Y^\circ, X_1)$ into the composition

$$\varphi\psi: Y^\circ \longrightarrow X_1 \longrightarrow X_2.$$

Clearly, $P_{\varphi\psi} = P_\varphi P_\psi$.

(2) Similarly, define $P^X \in \mathsf{C}^*$ by setting

$$P^X(Y) := \mathrm{Hom}_{\mathsf{C}}(X, Y) \quad \text{for any } Y \in \mathsf{C};$$

to every morphism $\varphi: Y_1 \longrightarrow Y_2$, we assign the map of sets $P^X(Y_1) \longrightarrow P^X(Y_2)$ which sends $\psi: X \longrightarrow Y_1$ into the composition $\psi\varphi: X \longrightarrow Y_1 \longrightarrow Y_2$.

To any $\varphi \in \mathrm{Hom}_{\mathsf{C}}(X_1, X_2)$, there corresponds a functor morphism $P^\varphi: P^{X_2} \longrightarrow P^{X_1}$ which to any $Y \in C$ assigns

$$P^\varphi(Y): P^{X_2}(Y) \longrightarrow P^{X_1}(Y)$$

and sends every morphism $\psi \in \mathrm{Hom}_{\mathsf{C}}(X_2, Y)$ into the composition

$$\psi\varphi: X_1 \longrightarrow X_2 \longrightarrow Y. \tag{1.33}$$

Clearly, $P^{\varphi\psi} = P^\psi P^\varphi$.

A functor $F: \mathsf{C}^\circ \longrightarrow$ **Sets** (resp. a functor $F: \mathsf{C} \longrightarrow$ **Sets**) is said to be *representable* (resp. *corepresentable*) if it is isomorphic to a functor of the form P_X (resp. P^X) for some $X \in \mathsf{C}$; then X is called an *object that represents F*.

1.16.8a Theorem

(1) *The map $\varphi \longmapsto P_\varphi$ defines an isomorphism of sets*

$$\mathrm{Hom}_{\mathbf{C}}(X, Y) \cong \mathrm{Hom}_{\mathbf{C}^*}(P_X, P_Y). \tag{1.34}$$

This isomorphism is functorial in both X and Y. Hence, the functor $P: \mathbf{C} \longrightarrow \mathbf{C}^*$ determines an equivalence of $\mathbf{C}$ with the full subcategory of $\mathbf{C}^*$ consisting of the representable functors.

(2) *If a functor of $\mathbf{C}^*$ is representable, the object that represents it is determined uniquely up to an isomorphism.*

Theorem 1.16.8a is the source of several important ideas.

1.16.8b The First Direction

It is convenient to think of P^X as of "the sets of points of an object $X \in \mathbf{C}$ *with values in various objects $Y \in \mathbf{C}$, or Y-points*"; notation: $P_X(Y)$, or more often $\underline{X}(Y)$. (The sets $P^X(Y)$ are also sometimes denoted by $\underline{X}(Y)$.)

In other words, $P_X = \bigsqcup_{Y \in \mathbf{C}} P_X(Y)$ with an additional structure: the sets of maps $P_X(Y_1) \longrightarrow P_X(Y_2)$ induced by morphisms $Y_2^\circ \longrightarrow Y_1^\circ$ for any $Y_1, Y_2 \in \mathbf{C}$ and compatible in the natural sense (the composition goes into the composition, and so on).

> **Therefore, in principle it is always possible to pass from
> the categorical point of view to the structural one, because
> all categorical properties of X are precisely mirrored
> by the categorical properties of the structure of P_X.
> The situation with P^X is similar.**

1.16.8c Remark (Motivation) Let $*$ be a one-point set. For categories with sufficiently "simple" structure of their objects, such as the category of finite sets or even the category of smooth finite-dimensional manifolds, $X = P_X(*)$ for every object X, i.e., X is completely determined by its $*$-points, i.e., just points.

For varieties (or for supermanifolds, *D.L.*), when the object may have either "sharp corners" or "inner degrees of freedom", the structure sheaf may contain nilpotents or zero divisors, so in order to keep this information and be able to completely describe X, we need *various types of points*, in particular, Y-points for some more complicated Y's.

1.16.8d The Second Direction

Replacing X by P_X (resp. by P^X) we can transfer conventional set-theoretical constructions to any category: an object $X \in \mathbf{C}$ is a *group, ring*, and so on *in the category* C, if the corresponding structure is given on every set $P_X(Y)$ of its Y-points and is compatible with the maps induced by the morphisms $Y_2^\circ \longrightarrow Y_1^\circ$ (resp. $Y_1 \longrightarrow Y_2$).[32]

1.16.8e   The Third Direction

Let C be a category of structures of a given type. Among the functors $C^\circ \longrightarrow$ Sets, i.e., among the objects of C^*, some natural functors may exist which a priori are constructed not as P_X or P^X, but eventually turn out to be representable. (*Examples*: the functor $X \longmapsto H^*(X; G)$ on the homotopy category.)

In such cases, it often turns out that the properties of the functor representable by such an object are in fact the most important properties of the object itself, which are obscured by its structural description.

It may well happen that some natural functors $C \longrightarrow$ Sets are not representable, though it is highly desirable that they would have been. The most frequent occurrence of such a situation is when we try to generalize to C some set-theoretical construction, e.g., factorization modulo a group action or modulo a more general relation. **In such a case it may help to add to C, considered as being embedded into C^*, the appropriate functors.**

In the algebraic geometry of 1970s this way of reasoning lead to monstrous creatures which B. Moishezon called *minischemes*, M. Artin *étale schemes*, and A. Grothendieck just *varietées*,[33] and later to *stacks*.

1.16.9   Proof of Theorem 1.16.8

Let us construct a map

$$i: \mathrm{Hom}_{C^*}(P_X, P_Y) \longrightarrow \mathrm{Hom}_C(X, Y)$$

which to every functor morphism $P_X \longrightarrow P_Y$ assigns the image of $\mathrm{id}_X \in P_X(X)$ in $P_Y(X)$ under the map $P_X(X) \longrightarrow P_Y(X)$ defined by this functor morphism. Let us verify that i and $\varphi \longmapsto P_\varphi$ are mutually inverse.

(1) $i(P_\varphi) = P_\varphi(\mathrm{id}_X) = \varphi$ by the definition of P_φ.

---

[32]This is exactly the way supergroups are defined and superalgebras *should be* defined. However, it is possible to define superalgebras using just one set-theoretical model, but sometimes we have to pay for this deceiving simplicity by wondering how to describe deformations with odd parameters.

[33]In supermanifold theory, this is one of the ways to come to *point-less* (or, perhaps more politely, *point-free*) supermanifolds, cf. [We].

(2) Conversely, given a functor morphism $g: P_X \longrightarrow P_Y$, we have by definition the maps $g(Z): P_X(Z) \longrightarrow P_Y(Z)$ for all $Z \in \mathrm{Ob}\,\mathbf{C}$. By definition, $i(g) = g(X)(\mathrm{id}_X)$ and we have to verify that $P_{i(g)}(Z) = g(Z)$.

By definition, $P_{i(g)}(Z)$ assigns to every morphism $g: Z \longrightarrow X$ the composition $i(g) \cdot \varphi: Z \longrightarrow X \longrightarrow Y$, and therefore we have to establish that $g(Z)(\varphi) = i(g) \cdot \varphi$. This follows from the commutativity of the diagram

$$
\begin{array}{ccc}
P_X(X) & \xrightarrow{\;g(x)\;} & P_Y(X) \\
{\scriptstyle P_X(\varphi)}\downarrow & & \downarrow{\scriptstyle P_Y(\varphi)} \\
P_X(Z) & \xrightarrow{\;g(z)\;} & P_Y(Z)
\end{array}
$$

Thus, we have verified that $\mathrm{Im}\,P$ is a full subcategory of $\mathbf{C}^*$, and therefore is equivalent to $\mathbf{C}$. The remaining statements are easily verified. $\square$

It is worth mentioning that if we add representable functors to the full subcategory of functors P_X of $\mathbf{C}^*$, i.e., if we add functors isomorphic to the ones the category already possesses, we obtain an equivalent subcategory.

1.16.10   The Object of Inner Homomorphisms in a Category

All morphisms of a set into another set constitute a set; morphisms of an Abelian group into another Abelian group constitute an Abelian group; there are many more similar examples. A natural way to axiomatize the situation when for $X, Y \in \mathrm{Ob}\,\mathbf{C}$ there is an *object of inner homomorphisms* $\underline{\mathrm{Hom}}(Y, X) \in \mathrm{Ob}\,\mathbf{C}$, is to define the corresponding representable functor $\underline{\mathrm{Hom}}$. Quite often it is determined by the formula

$$
\mathrm{Hom}_{\mathbf{C}}(X, \underline{\mathrm{Hom}}(Y, Z)) = \mathrm{Hom}_{\mathbf{C}}(X * Y, Z) \tag{1.35}
$$

for an appropriate operation $*$.[34]

---

[34] As examples of categories with interesting objects of inner homomorphisms, we can take any of the categories of superspaces, or superalgebras, or supermanifolds, or supergroups.

1.17 Solutions to Selected Problems of Chap. 1

1.17.1 Exercise 1.3.4

(1) Since $\mathfrak{a}_1 \cdots \mathfrak{a}_n \subset \mathfrak{a}_1 \cup \ldots \cup \mathfrak{a}_n$, it follows that $V(\mathfrak{a}_1 \cdots \mathfrak{a}_n) \supset V(\mathfrak{a}_1 \cup \ldots \cup \mathfrak{a}_n)$. The other way round, let $\mathfrak{p}_x$ be a prime ideal. Then

$$\mathfrak{p}_x \in V(\mathfrak{a}_1 \cdots \mathfrak{a}_n) \Longleftrightarrow x \in \bigcap_i V(\mathfrak{a}_i) \text{ means that } x \in V(\mathfrak{a}_{i_0}) \text{ for some } i_0,$$

hence, $\mathfrak{p}_x \supset \mathfrak{a}_{i_0} \supset \mathfrak{a}_1 \cup \cdots \cup \mathfrak{a}_n$ for this i_0, i.e., $x \in V(\mathfrak{a}_1 \cup \cdots \cup \mathfrak{a}_n)$.

(2) By Lemma 1.4.1, we have

$$(g_1, \ldots, g_n) = A \Longleftrightarrow \bigcap_{i=1}^{n} V(g_i) = \emptyset.$$

Moreover, $V(g^m) = V(g)$ for $m > 0$, implying the desired result.

1.17.2 Exercise 1.6.7

(3) Let f/g, where $f, g \in A$, be such that

$$(f/g)^n + a_{n-1}(f/g)^{n-1} + \cdots + a_0 = 0, \text{ where } a_i \in A.$$

Then $f^n + f a_{n-1} f^{n-1} + \cdots + g^n a_0 = 0$, implying $g \mid f^n$, and if $(g, f) = 1$, then g is invertible in A. Therefore, $f/g \in A$.

Chapter 2
Sheaves, Schemes, and Projective Spaces

2.1 Basics on Sheaves

The topological space $\operatorname{Spec} A$ is by itself a rather coarse invariant of A, see Examples 1.5.3. Therefore, as the "right" geometric object corresponding to A, it is natural to take the pair $(\operatorname{Spec} A, \widetilde{A})$ consisting of the space $\operatorname{Spec} A$ and the set of elements of A considered, more or less adequately, as functions on $\operatorname{Spec} A$, and so we did up to now.

But, on the other hand, only *local* geometric objects are associated with rings; so in order to construct, say, a projective space, we have to glue it from affine spaces. Let us learn how to do so.

The gluing procedure we are interested in may be described, topologically, as follows: let X_1, X_2 be two topological spaces, $U_1 \subset X_1$ and $U_2 \subset X_2$ open subsets, and $f\colon U_1 \longrightarrow U_2$ an isomorphism of some kind. Then we construct the set

$$X = (X_1 \cup X_2)/R_f,$$

where R_f is the equivalence relation that identifies the points that correspond to each other under f. On X, a natural topology is induced, and we say that X *is the result of gluing X_1 with X_2 by means of f.*

Figure 2.1 illustrates two ways to glue two affine lines, $X_1 = X_2 = \mathbb{R}$, with $U_1 = U_2 = \mathbb{R} \setminus \{0\}$. The results are:

(a) The line with a double point (here $f = \operatorname{id}$).
(b) $\mathbb{P}^1$, the projective line (here $f(x) = x^{-1}$). Clearly, $\mathbb{P}^1$ is homeomorphic to the circle S^1.

When we try to apply this construction to spectra of rings we immediately encounter the above-mentioned circumstance, namely, that the topological structure of the open sets insufficiently reflects the algebraic information which we would like

© The Author(s) 2018
Y.I. Manin, *Introduction to the Theory of Schemes*, Moscow Lectures 1,
https://doi.org/10.1007/978-3-319-74316-5_2

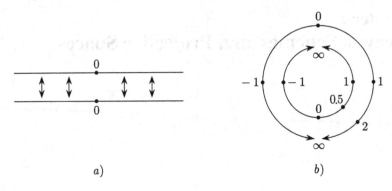

a)   b)

Fig. 2.1

to preserve and which is carried by the ring A itself. The theories of differentiable manifolds and analytic varieties suggest a solution.

In order to glue a differentiable manifold from two balls X_1 and X_2, we require that the isomorphism $f: X_1 \longrightarrow X_2$ that affects the gluing be not just an isomorphism of topological spaces, but also preserve the differentiable structure. This means that the map f^*, which sends continuous functions on X_2 into continuous functions on X_1, must induce an isomorphism of subrings of differentiable functions, otherwise the gluing would not be "smooth".

The analytic case is similar.

Therefore, we have to consider functions of a certain class which are defined on various open subsets of the topological space X.

The relations between functions defined on different open sets are axiomatized by the following definition.

2.1.1 Presheaves

Fix a topological space X. Let $\mathcal{P}$ be a rule that to every open set $U \subset X$ assigns a set $\mathcal{P}(U)$ and, for any pair of open subsets $U \subset V$, assigns a *restriction map* $r_U^V: \mathcal{P}(V) \longrightarrow \mathcal{P}(U)$ such that

(1) $\mathcal{P}(\emptyset)$ consists of one element;
(2) $r_U^W = r_U^V \circ r_V^W$ for any open subsets (briefly: *opens*) $U \subset V \subset W$.

Then the collection $\{\mathcal{P}(U), r_U^V \mid U, V \text{ opens in } X\}$ is called a *presheaf* (of sets) on X.

The elements of the set $\mathcal{P}(U)$, which is often denoted by $\Gamma(U, \mathcal{P})$, are called the *sections* of the presheaf $\mathcal{P}$ over U; a section can be regarded as a "function" defined over U.

2.1.1a Remark Axiom (1) is convenient in some subtle considerations of category theory. Axiom (2) expresses the natural transitivity of restriction.

2.1.2 The Category Top$_X$

The objects of Top$_X$ are the open subsets of X and its morphisms are inclusions. Thus, a *presheaf* of sets on X is a functor $\mathcal{P}\colon \text{Top}_X^\circ \longrightarrow \text{Sets}$.

Genuine functions can be multiplied, added and multiplied by scalars; accordingly, we can consider presheaves of groups, rings, and so on. A formal definition is as follows:

Let $\mathcal{P}$ be a presheaf of sets on X; if every set $\mathcal{P}(U)$ carries an algebraic structure (of a group, ring, A-algebra, and so on) and the restriction maps r_U^V are homomorphisms of this structure, i.e., $\mathcal{P}$ is a functor $\text{Top}_X^\circ \longrightarrow \text{Gr}$ (Ab, Rings, A-Algs, and so on), then $\mathcal{P}$ is called a *presheaf of groups, Abelian groups, rings, A-algebras,* and so on, respectively.

Finally, we may consider external composition laws, e.g., a presheaf of modules over a presheaf of rings (given on the same topological space).

2.1.2a Exercise Give a formal definition of such external composition laws.

2.1.3 Sheaves

The presheaves of continuous (infinitely differentiable, analytic, and so on) functions on a space X possess additional properties (of "analytic continuation" type) which are axiomatized in the following definition.

A presheaf $\mathcal{P}$ on a topological space X is called a *sheaf* if it satisfies the following condition: For any open subset $U \subset X$, any open cover $U = \bigcup_{i \in I} U_i$, and any collection of sections $s_i \in \mathcal{P}(U_i)$, where $i \in I$, such that

$$r_{U_i \cap U_j}^{U_i}(s_i) = r_{U_i \cap U_j}^{U_j}(s_j) \text{ for all } i, j \in I,$$

there exists a section $s \in \mathcal{P}(U)$ such that $s_i = r_{U_i}^U(s)$ for all $i \in I$, and such a section s is unique.

In other words, from a set of compatible sections over the opens U_i one can glue a section over U, and any section over U is uniquely determined by the collection of its restrictions to the sets U_i.

2.1.3a Remark If $\mathcal{P}$ is a presheaf of Abelian groups, the following reformulation of the above condition is useful:

A presheaf $\mathcal{P}$ is a sheaf if, for any $U = \bigcup_{i \in I} U_i$, the following sequence of Abelian groups is exact:

$$0 \longrightarrow \mathcal{P}(U) \xrightarrow{\varphi} \prod_{i \in I} \mathcal{P}(U_i) \xrightarrow{\psi} \prod_{i,j \in I} \mathcal{P}(U_i \cap U_j),$$

where φ and ψ are given by the formulas

$$\varphi(s) = (\ldots, r_{U_i}^U(s), \ldots)$$
$$\psi(\ldots, s_i, \ldots, s_j, \ldots) = (\ldots, r_{U_i \cap U_j}^{U_i}(s_i) - r_{U_i \cap U_j}^{U_j}(s_j), \ldots).$$

For a general presheaf of Abelian groups, one can assert only that this sequence is a complex. (Its natural extension determines a Čech cochain complex that we will study in what follows.)

2.1.4   The Relation Between Sheaves and Presheaves

The natural objects are sheaves, but various constructions with them often lead to presheaves which are not sheaves.

2.1.4a Example (This example is of a fundamental importance for cohomology theory.) Let $\mathcal{F}_1$ and $\mathcal{F}_2$ be two sheaves of Abelian groups and $\mathcal{F}_1 \subset \mathcal{F}_2$, i.e., $\mathcal{F}_1(U) \subset \mathcal{F}_2(U)$, with compatible restriction maps. As is easy to see, the collection of groups $\mathcal{P}(U) = \mathcal{F}_1(U)/\mathcal{F}_2(U)$ and natural restriction maps is a presheaf but, in general, not a sheaf. Here is a typical example:

Let X be a circle, $\mathcal{O}$ the sheaf over X for which $\mathcal{O}(U)$ is the group of continuous $\mathbb{R}$-valued functions on U, and $\widetilde{\mathbb{Z}} \subset \mathcal{O}$ the "constant" presheaf, i.e., $\widetilde{\mathbb{Z}}(U) = \mathbb{Z}$ for each non-empty U. The presheaf $\mathcal{P}$ for which

$$\mathcal{P}(U) = \mathcal{O}(U)/\widetilde{\mathbb{Z}}(U) \text{ for every open subset } U \subset X$$

is not a sheaf, for the following reason.

Consider two connected open sets $U_1, U_2 \subset X$ such that $X = U_1 \cup U_2$ and $U_1 \cap U_2$ is the union of two connected components V_1, V_1 (e.g., U_1, U_2 are slightly enlarged half-circles). Let $f_1 \in \mathcal{O}(U_1), f_2 \in \mathcal{O}(U_2)$ be two continuous functions such that

$$r_{V_1}^{U_1}(f_1) - r_{V_1}^{U_2}(f_2) = 0, \quad r_{V_2}^{U_1}(f_1) - r_{V_2}^{U_2}(f_2) = 1.$$

Then the classes $f_1 \bmod \mathbb{Z} \in \mathcal{O}(U_1)/\widetilde{\mathbb{Z}}(U_1)$ and $f_2 \bmod \mathbb{Z} \in \mathcal{O}(U_2)/\widetilde{\mathbb{Z}}(U_2)$ agree over $U_1 \cap U_2$.

On the other hand, the sheaf $\mathcal{O}$ over X has no section whose restrictions to U_1 and U_2 are $f_1 \bmod \mathbb{Z}$ and $f_2 \bmod \mathbb{Z}$, respectively, because it is impossible to remove the incompatibility on V_2. The cause is, clearly, the fact that the circle is not simply-connected (Fig. 2.2).

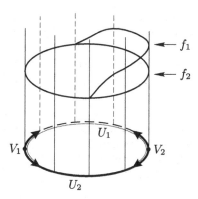

Fig. 2.2

2.1.5   Gluing Sheaves from Presheaves

With every presheaf $\mathcal{P}$ there is associated in a canonical manner a certain sheaf $\mathcal{P}^+$. The sections of $\mathcal{P}^+$ are constructed from the sections of $\mathcal{P}$ by means of two processes.

The first process *reduces* the number of sections in $\mathcal{P}(U)$ identifying those which begin to coincide once restricted to a sufficiently fine cover $U = \bigcup_i U_i$.

The second process *increases* the number of sections in $\mathcal{P}(U)$ by adding the sections glued from the compatible sets of sections on the covers of U.

Clearly, a passage to limit is required. A notion that is technically convenient here, as well as in other problems, is the following one. Let $\mathcal{P}$ be a presheaf over X. The *stalk* $\mathcal{P}_x$ of $\mathcal{P}$ at a point $x \in X$ is the set $\varinjlim \mathcal{P}(U)$, where the inductive limit[1] is taken with respect to the directed system of open neighborhoods of x.

---

[1]Recall the definition, see [StE]. Let the objects of a category be sets with algebraic structures (such as groups, rings, modules (over a fixed ring), and so on) and its morphisms be the morphisms of these structures. Let $(I, \leq)$ be an inductive set. Let $\{A_i \mid i \in I\}$ be a family of objects labelled by the elements of I, and for all $i \leq j$, let a family of homomorphisms $f_{ij}\colon A_i \longrightarrow A_j$ be given having the following properties:

(a) $f_{ii} = \mathrm{id}\,|_{\{A_i\}}$;
(b) $f_{ik} = f_{jk} \circ f_{ij}$ for all $i \leq j \leq k$.

Then the pair (A_i, f_{ij}) is said to be *directed (inductive) system over* I. The *inductive* (or *direct*) *limit* A of the inductive system (A_i, f_{ij}) is defined as the quotient of the disjoint union of the sets A_i modulo an equivalence relation $\sim$:

$$\varinjlim A_i = \coprod_i A_i \big/ \sim,$$

were $\sim$ is defined as follows: if $x_i \in A_i$ and $x_j \in A_j$, then $x_i \sim x_j$ if, for some $k \in I$, we have $f_{ik}(x_i) = f_{jk}(x_j)$.

The elements of $\mathcal{P}_x$ are called *germs* of sections over x. A *germ* is an equivalence class in the set of sections over different open neighborhoods U containing x modulo the equivalence relation which identifies sections $s_1 \in \mathcal{P}(U_1)$ and $s_2 \in \mathcal{P}(U_2)$ whenever their restrictions to a common subset $U_3 \subset U_1 \cap U_2$ containing x coincide.

For any point x and an open neighborhood $U \ni x$, one has a naturally defined map $r_x^U \colon \mathcal{P}(U) \longrightarrow \mathcal{P}_x$.

Clearly, $\mathcal{P}_x$ carries the same structure as the sets $\mathcal{P}(U)$; i.e., it is a group, ring, module, and so on, when $\mathcal{P}$ is a presheaf of groups, rings, modules, and so on.

The idea underlying the construction of the sheaf $\mathcal{P}^+$ from a given presheaf $\mathcal{P}$ is that the sections of $\mathcal{P}^+$, i.e., elements of $\mathcal{P}^+(U)$, are defined as certain sets of germs of sections, i.e., as elements of $\prod_{x \in U} \mathcal{P}_x$ that are compatible in the following natural sense.

Let $\mathcal{P}$ be a presheaf on a topological space X. For every nonempty open subset $U \subset X$, define the subset $\mathcal{P}^+(U) \subset \prod_{x \in U} \mathcal{P}_x$ by

$$
\mathcal{P}^+(U) = \left\{
\begin{array}{l}
(\ldots, s_x, \ldots) \mid \text{ for any } x \in U, \text{ there exist} \\
\text{an open neighborhood } V \subset U \text{ of } X, \text{ and a section } s \in \mathcal{P}(V) \\
\text{such that } s_y = r_y^V(s) \text{ for all } y \in V.
\end{array}
\right\}
$$

Further, for any pair of open subsets $V \subset U$, define the restriction map $\mathcal{P}^+(U) \longrightarrow \mathcal{P}^+(V)$ as the one induced by the projection $\prod_{x \in U} \mathcal{P}_x \longrightarrow \prod_{x \in V} \mathcal{P}_x$.

2.1.6 Theorem *The family of sets $\mathcal{P}^+(U)$ together with the described restriction maps is a sheaf on X, and $\mathcal{P}_x^+ = \mathcal{P}_x$ for any $x \in X$.*

The sheaf $\mathcal{P}^+$ is called the *sheaf associated with* $\mathcal{P}$. Clearly, the algebraic structures are transplanted from $\mathcal{P}$ to $\mathcal{P}^+$.

Proof is a straightforward verification of definitions and is left to the reader.

2.1.7 Another Definition of Sheaves

While the definition given below is equivalent to the previous one, given in Sect. 2.1.3, it is sometimes more convenient to grasp intuitively.

A *sheaf* $\mathcal{F}$ on a topological space X is a pair $(Y_{\mathcal{F}}, r)$, where $Y_{\mathcal{F}}$ is a topological space and $r \colon Y_{\mathcal{F}} \longrightarrow X$ is an open continuous map to X such that, for any $y \in Y_{\mathcal{F}}$, there exists an open neighborhood of y in $Y_{\mathcal{F}}$ such that r is a local homeomorphism in this neighborhood.

The relation between this definition and the previous one is as follows. Given $r \colon Y_{\mathcal{F}} \longrightarrow X$, we define a sheaf as in Sect. 2.1.3 by setting $\mathcal{F}(U)$ to be the set of local sections of $Y_{\mathcal{F}}$ over U, i.e., the maps $s \colon U \longrightarrow Y_{\mathcal{F}}$ such that $r \cdot s = \mathrm{id}\,|_U$.

Conversely, if $\mathcal{F}$ is given by its sections over U for all $U \subset X$, set

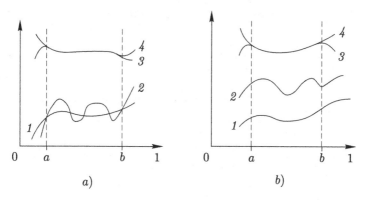

Fig. 2.3

$$Y_{\mathcal{F}} = \bigcup_{x \in X} \mathcal{F}_x,$$

and let r be the map $r \colon \mathcal{F}_x \longrightarrow x$. Define the topology on $Y_{\mathcal{F}}$ by declaring all sections $s \in \mathcal{F}(U)$ as open sets, i.e., put

$$s = \bigcup_{x \in U} r_x^U(s) \subset \bigcup_{x \in U} \mathcal{F}_x \subset \bigcup_{x \in X} \mathcal{F}_x$$

One should bear in mind that, even for Hausdorff spaces X, the spaces $Y_{\mathcal{F}}$ are not necessarily Hausdorff.

Figure 2.3 depicts a part of the space $Y_{\mathcal{F}}$ corresponding to the sheaf of continuous functions on the segment $[0, 1]$. To the graph of every continuous function there corresponds an open subset of $Y_{\mathcal{F}}$, and the sections corresponding to the graphs of functions do not intersect in $Y_{\mathcal{F}}$ unless the graphs intersect.

If the functions f_1, f_2 coincide over a (necessarily *closed*!) set Y, then the corresponding sections of $Y_{\mathcal{F}}$ coincide over the *interior* of Y. In the space $Y_{\mathcal{F}}$, the points over a (or b) belonging to sections 3 and 4 are distinct, but any two neighborhoods of these points intersect.   □

2.2   The Structure Sheaf on Spec A

We consider the elements of A as functions on Spec A. Now, if we wish to consider restrictions of these functions, we need to construct from them functions with smaller domains. The only such process which does not require limits is to consider fractions f/g, because $f(x)/g(x)$ makes sense for all $x \notin V(g)$. This suggests the following definition.

2.2.1   The Case of Rings Without Zero Divisors

Let A be a ring without zero divisors, $X = \operatorname{Spec} A$, K the field of fractions of A. For any nonempty open subset $U \subset X$, denote

$$\mathcal{O}_X(U) = \{a \in K \mid a \text{ can be represented in the form } f/g,$$
$$\text{where } g(x) \neq 0 \text{ for any } x \in U\}.$$

If $U \subset V$, we denote the obvious embedding $\mathcal{O}_X(V) \hookrightarrow \mathcal{O}_X(U)$ by r_U^V.

2.2.1a Theorem

(1) The presheaf $\mathcal{O}_X$ defined above is a sheaf of rings.
(2) The stalk of $\mathcal{O}_X$ at x is

$$\mathcal{O}_x = \{f/g \mid f, g \in A,\ g \notin \mathfrak{p}_x\},$$

and for any nonzero $g \in A$, we have

$$\Gamma(D(g), \mathcal{O}_X) = \{f/g^n \mid f \in A,\ n \geq 0\}.$$

2.2.1b Remark Assertion (2) of Theorem 2.2.1a is less trivial. Intuitively it reflects two things:

(2a) If an "algebraic function" is defined at points where $g \neq 0$, then the worst that may happen to it on $V(g)$ is a pole of finite order: there are no essentially singular points;
(2b) Over big open sets, i.e., sets of the form $D(g)$, there is no need to consider "glued" sections: they are given by a single equation.

On the other hand, the description of the stalks $\mathcal{O}_x$ enables us to identify $\operatorname{Spec} A$ with a collection of subrings of the field K. This is the point of view pursued in works by Chevalley and Nagata.

Proof The fact that $\mathcal{O}_X$ is a sheaf is established by trivial reference to definitions, and the same for $\mathcal{O}_x$. The reason for this is that all stalks $\mathcal{O}_x$ are embedded into K, so that the relations of "extension" and "restriction" are induced by the identity relation in K. (In what follows, we give a longer definition suitable for any rings, even when K, the localization of A, does not exist.)

Here we prove assertion (2) only for $g = 1$, just to illustrate the main idea. The general case is treated in the following subsection.

Thus, we would like to show that

if an element of the field of fractions K can be represented as a rational fraction whose denominator does not belong to any prime ideal of A, then this element necessarily belongs to A, i.e., there is no denominator at all.

This is obvious for unique factorization rings.

In the general case, the argument (due to Serre) is suggested by an analogy with differentiable manifolds and uses "partition of unity".

Let $f \in \Gamma(X, \mathcal{O}_X) \subset K$. For any point $x \in \operatorname{Spec} A$, set $f = g_x/h_x$, where g_x, $h_x \in A$ and $h_x \notin \mathfrak{p}_x$. Let $U_x = D(h_x)$. Obviously, U_x is an open neighborhood of x.

Now, apply Proposition 1.4.13 on quasi-compactness to the cover $X = \bigcup_x D(h_x)$.

Let $x_i \in X$, and $a_i \in A$ for $i = 1, \ldots, n$. Set $h_i = h_{x_i}$ and $g_i = g_{x_i}$. Let $1 = \sum_{1 \le i \le n} a_i h_i$ in A. Then $X = \bigcup_{1 \le i \le n} D(h_i)$ and

$$f = \sum_{1 \le i \le n} a_i h_i f = \sum_{1 \le i \le n} a_i h_i (g_i h_i) = \sum_{1 \le i \le n} a_i g_i \in A. \qquad \square$$

2.2.2 Example of Calculation of $\mathcal{O}_X(U)$ for Opens Other Than the Big Ones

In $\operatorname{Spec} K[T_1, T_2]$, where K is a field, let $U = D(T_1) \cup D(T_2)$, i.e., U is the complement of the origin. Since $K[T_1, T_2]$ has no zero divisors, we have

$$\mathcal{O}_X(U) = K[T_1, T_2, T_1^{-1}] \cap K[T_1, T_2, T_2^{-1}].$$

By the unique factorization property in polynomial rings we immediately see that $\mathcal{O}_X(U) = K[T_1, T_2]$, and therefore a function on the plane cannot have a singularity supported at a closed point: the function is automatically defined at it.

Similar arguments are applicable to a multidimensional affine space: if $n \ge 2$ and $F_1, \ldots, F_n \in K[T_1, \ldots, T_n]$ are relatively prime, then

$$\mathcal{O}_X\left(\bigcup_{1 \le i \le n} D(F_i) \right) = K[T_1, \ldots, T_n].$$

The support of the set of singularities of a rational function cannot be defined by more than essentially one equation.

2.2.3 The Structure Sheaf on Spec A: General Case

If A has zero divisors, then its field of fractions does not exist. Therefore, the algebraic formalism necessary for a correct definition of rings of fractions and relations between them becomes more involved. Nevertheless, sheaves are introduced in an essentially the same way as for the rings without zero divisors, and with the same results.

Constructing a sheaf on $\operatorname{Spec} A$ we have to study the dependence of A_S on S, and the ring homomorphisms $A_S \longrightarrow A_{S'}$ for different S and S'. Theorem 1.6.4c stating the universal character of localization, plays a fundamental role in what follows.

2.2.4   The Structure Sheaf $\mathcal{O}_X$ on $X = \operatorname{Spec} A$

For every $x \in X$, set (see Sect. 1.6.2)

$$\mathcal{O}_x := A_{A \setminus \mathfrak{p}_x} = A_{\mathfrak{p}_x}.$$

For any open subset $U \subset X$, define the ring of sections of the presheaf $\mathcal{O}_X$ over U to be the subring

$$\mathcal{O}_X(U) \subset \prod_{x \in U} \mathcal{O}_x$$

consisting of the elements $(\ldots, s_x, \ldots)$, where $s_x \in \mathcal{O}_x$, with the property that for every point $x \in U$, there exist a big open set, a neighborhood $D(f_x)$ of x for some $f_x \in A$, and an element $g \in A_{f_x}$, such that s_y is the image of g under the natural homomorphism $A_{f_x} \longrightarrow \mathcal{O}_y$ for all $y \in U$.

Define the restriction maps r_U^V as the homomorphisms induced by the projections $\prod_{x \in V} \mathcal{O}_x \longrightarrow \prod_{x \in U} \mathcal{O}_x$. It is easy to see that $\mathcal{O}_X$ is well defined and the natural homomorphism $A_{f_x} \longrightarrow \mathcal{O}_y$ is induced by the embedding of multiplicative sets

$$\{f_x^n \mid n \in \mathbb{Z}_+\} \subset A \setminus \mathfrak{p}_y.$$

2.2.4a Theorem *The presheaf $\mathcal{O}_X$ is a sheaf whose stalk over $x \in X$ is isomorphic to $\mathcal{O}_x$ and in which r_x^U is the composition*

$$\mathcal{O}_X(U) \longrightarrow \prod_{x' \in U} \mathcal{O}_{x'} \xrightarrow{\text{pr}} \mathcal{O}_x$$

Furthermore, the ring homomorphism

$$j \colon A_f \longrightarrow \mathcal{O}_X(D(f)), \quad j(g/f) = (\ldots, j_x(g/f), \ldots)_{x \in U},$$

where $j_x \colon A_f \longrightarrow \mathcal{O}_x$ is a natural homomorphism of rings of fractions, is an isomorphism.

The sheaf $\mathcal{O}_X$ on the scheme $X = \operatorname{Spec} A$ is called the *structure sheaf* of X.

Proof The fact that $\mathcal{O}_X$ is a sheaf follows immediately from the definitions and the compatibility of natural homomorphisms of rings of fractions. By definition, the stalk $\mathcal{O}_{X,x}$ of $\mathcal{O}_X$ at x is equal to

$$\varinjlim \mathcal{O}_X(U) = \varinjlim_{f(x)\neq 0} \mathcal{O}_X(D(f)),$$

because the sets $D(f)$ constitute a basis of neighborhoods of x. The natural homomorphisms $A_f \longrightarrow \mathcal{O}_X(D(f)) \longrightarrow \mathcal{O}_x$ determine the homomorphism $\mathcal{O}_{X,x} \longrightarrow \mathcal{O}_x$. It is an epimorphism, because any element of $\mathcal{O}_x$ is of the form g/f, where $f(x) \neq 0$, and therefore is the image of the corresponding element of $\mathcal{O}_X(D(f))$.

Moreover, the kernel of this homomorphism is trivial: if g/f goes into $0 \in \mathcal{O}_x$, then by Lemma 1.6.4b $f_1 g = 0$ for some f_1 such that $f_1(x) \neq 0$, and therefore the image of g/f in A_{ff_1}, and the more so in the inductive limit $\varinjlim \mathcal{O}_X(D(f))$, is equal to 0.

It remains to establish the "furthermore" part of the theorem; notice that it gives a "finite" description of the rather cumbersome and bulky rings $\mathcal{O}_X(U)$ for big open sets U.

First of all, $\operatorname{Ker} j = 0$.

Indeed, if $j(g/f) = 0$, then, for every point $x \in D(f)$, there exists an element $t_x \in \operatorname{Ann} g = \{a \mid ag = 0\}$ such that $t_x(x) \neq 0$. But this means that $\operatorname{Ann} g \not\subset \mathfrak{p}_x$, i.e., $x \notin V(\operatorname{Ann} g)$ for $x \in D(f)$. Therefore, $V(\operatorname{Ann} g) \subset V(f)$, i.e., $f^n \in \operatorname{Ann} g$ for some n. Hence, $g/f = 0$ in A_f.

Now, let us prove that j is an epimorphism. Let $s \in \mathcal{O}_X(D(f))$ be a section. By definition, there exists a cover $D(f) = \bigcup_{x \in D(f)} D(h_x)$ such that s is induced over $D(h_x)$ by an element g_x/h_x. As in Sect. 2.2.1, let us construct a partition of unity, or rather not of unity, but of an invertible on $D(f)$ function f^n:

$$V(f) = \bigcap_{x \in D(f)} V(h_x) = V\left(\bigcap_{x \in D(f)} \{h_x\} \right),$$

implying $f^n = \sum_{x \in D(f)} a_x h_x$. Since $a_x = 0$ for almost all x, we may denote the incontractible decomposition as

$$D(f) = \bigcup_{1 \leq i \leq r} D(h_i), \text{ where } f^n = \sum_{1 \leq i \leq r} a_i h_i \text{ with } a_i = a_{x_i}, h_i = h_{x_i}$$

Now, consider our section s glued together from the fractions g_i/h_i. The fact that g_i/h_i and g_j/h_j agree on $D(h_i) \cap D(h_j) = D(h_i h_j)$ means that the images of g_i/h_i and g_j/h_j coincide in all rings $\mathcal{O}_x$, where $x \in D(h_i h_j)$.

By what we proved above, $g_i/h_i - g_j/h_j = 0$ in $A_{h_i h_j}$, i.e., for some $m \in \mathbb{Z}_+$ (which may be chosen to be independent of the indices because the cover is finite) we have $(h_i h_j)^m (g_i h_j - g_j h_i) = 0$. Replacing h_i by h_i^{m+1} and g_i by $g_i h_i^m$ we may assume that $m = 0$.

Now, the compatibility conditions take the form $g_i h_j = g_j h_i$. Therefore,

$$f^n g_j = \sum a_i h_i g_j = \left(\sum_{1 \le i \le r} a_i g_i \right) h_j,$$

implying that the image of $\sum_{1 \le i \le r} a_i g_i / f^n$ in A_{h_j} is precisely g_j / h_j.

Thus, the compatible local sections over $D(h_j)$ are the restrictions of one element of A_f, as required.   □

The sheaf on $X = \operatorname{Spec} A$ described above will sometimes be denoted by $\widetilde{A}$. The pair $(\operatorname{Spec} A, \widetilde{A})$ consisting of a topological space and a sheaf over it determines the ring A thanks to Theorem 2.2.4: namely, $A = \Gamma(\operatorname{Spec} A, \widetilde{A})$. This pair is the main local object of algebraic geometry.

2.3 Ringed Spaces: Schemes

2.3.1 Ringed Spaces

A *ringed topological space* is a pair $(X, \mathcal{O}_X)$ consisting of a space X and a sheaf of rings $\mathcal{O}_X$ on it, called the *structure sheaf*.

A *morphism of ringed spaces* $F: (X_1, \mathcal{O}_X) \longrightarrow (Y_1, \mathcal{O}_Y)$ is a pair consisting of a morphism $f: X \longrightarrow Y$ of topological spaces and the collection of ring homomorphisms

$$f_U^*: \mathcal{O}_Y(U) \longrightarrow \mathcal{O}_X(f^{-1}(U)) \text{ for every open } U \subset Y$$

that are compatible with the restriction maps, i.e., such that

(a) the diagrams

$$
\begin{array}{ccc}
\mathcal{O}_Y(U) & \xrightarrow{\ f_U^*\ } & \mathcal{O}_X(f^{-1}(U)) \\
{\scriptstyle r_V^U} \downarrow & & \downarrow {\scriptstyle r_{f^{-1}(V)}^{f^{-1}(U)}} \\
\mathcal{O}_Y(V) & \xrightarrow{\ f_V^*\ } & \mathcal{O}_X(f^{-1}(V))
\end{array}
$$

commute for every pair of open subsets $V \subset U$ in Y;

(b) for any open set $U \subset Y$, and a pair $u \in U$ and $g \in \mathcal{O}_Y(U)$ such that $g(u) = 0$, we have

$$f_U^*(g)(x) = 0 \text{ for all } x \text{ such that } f(x) = u.$$

2.3.1a Remark *(Elucidation)* If X and Y are Hausdorff spaces, and $\mathcal{O}_X$ and $\mathcal{O}_Y$ are the sheaves of continuous (smooth, analytic, and so on) functions on them, respectively, then to every morphism $f: X \longrightarrow Y$ there corresponds the ring homomorphism $f_U^*: \mathcal{O}_Y(U) \longrightarrow \mathcal{O}_X(f^{-1}(U))$: namely, f_U^* assigns to every function $g \in \mathcal{O}_Y(U)$ the function

$$f_U^*(g)(x) = g(f(x)) \text{ for all } x \in f^{-1}(U);$$

i.e., the domain of $f_U^*(g)$ is $f^{-1}(U)$, and $f_U^*(g)$ is constant on the preimage of every $y \in U$.

In algebraic geometry, the spaces are not Hausdorff and their structure sheaves are not readily recognized as sheaves of functions. Therefore,

1. the collection of ring homomorphisms $\{f_U^* \mid U$ is an open set$\}$ is **not** recovered from f and *must be given separately*;
2. the condition

$$f_U^*(g)(x) = g(f(x)) \text{ for any } x \in f^{-1}(U)$$

is replaced by the weaker condition (b) above.

The reason for these two distinctions from the case of usual functions is that the domains of our "make believe" functions vary and different sections of the structure sheaf may represent the same function.

A ringed space isomorphic to one of the form $(\operatorname{Spec} A, \widetilde{A})$ is called an *affine scheme*. We will prove in next subsections the equivalence of this definition of affine scheme to that given earlier (Sect. 1.5.2).

2.3.2 Schemes

A ringed topological space $(X, \mathcal{O}_X)$ is called a *scheme* if every point x of it has an open neighborhood U such that $(U, \mathcal{O}_X|_U)$ is an affine scheme.

One of the methods for explicit description of a global object is to specify the local objects from which the object in question is glued and the gluing recipe. Here is the formal procedure.

2.3.2a Proposition *Let $(X_i, \mathcal{O}_{X_i})_{i \in I}$, be a family of schemes and suppose that in every X_i there are given open subsets U_{ij}, where $i, j \in I$. Further, suppose there is given a collection of isomorphisms*

$$\theta_{ij}: (U_{ij}, \mathcal{O}_{X_i}|_{U_{ij}}) \longrightarrow (U_{ji}, \mathcal{O}_{X_j}|_{U_{ji}})$$

satisfying the cocycle condition

$$\theta_{ii} = \text{id}, \quad \theta_{ij} \circ \theta_{ji} = \text{id}, \quad \theta_{ij} \circ \theta_{jk} \circ \theta_{ki} = \text{id}.$$

Then there exist a scheme $(X, \mathcal{O}_X)$, an open cover $X = \bigcup_{i \in I} X_i'$, and a family of isomorphisms $\varphi_i \colon (X_i', \mathcal{O}_X|_{X_i'}) \longrightarrow (X_i, \mathcal{O}_{X_i})$, such that

$$(\varphi_j|_{X_i \cap X_j})^{-1} \circ \theta_{ij} \circ \varphi_i|_{X_i \cap X_j} = \text{id} \ \text{for all } i, j.$$

2.3.2b Exercise Prove this proposition.

Hint For any open subset U, the space $(U, \mathcal{O}_X|_U)$ is also a scheme. Indeed, let $x \in U$; then x has a neighborhood $U_x \subset X$ such that $(U_x, \mathcal{O}_X|_{U_x})$ is isomorphic to $(\text{Spec}\, A, \widetilde{A})$. Then, $U \cap U_x$ is a nonempty open subset of $\text{Spec}\, A$, and because the big open sets $D(f)$, where $f \in A$, constitute a basis of the Zariski topology and $(D(f), \widetilde{A}|_{D(f)}) \cong (\text{Spec}\, A_f, \widetilde{A}_f)$, we can find an affine neighborhood of x that lies inside U.

2.3.3 Examples

In the examples below, $(X_i, \mathcal{O}_{X_i})$ will most often be affine schemes.

2.3.3a. Projective Spaces

Let K be a ring. We define the scheme $\mathbb{P}_K^n$, the n-dimensional projective space over K. Let $T_0, \dots, T_n$ be indeterminates. Set

$$U_i = \text{Spec}\, K\left[\frac{T_0}{T_i}, \dots, \frac{T_n}{T_i}\right], \quad U_{ij} = \text{Spec}\, K\left[\frac{T_0}{T_i}, \dots, \frac{T_n}{T_i}\right]_{T_j/T_i} \subset U_i.$$

Introduce the scheme isomorphism

$$\theta_{ij} : U_{ij} \longrightarrow U_{ji}$$

by naturally identifying the corresponding rings of fractions with the ring whose elements are of the shape $\dfrac{f(T_0, \dots, T_n)}{T_i^a T_j^b}$, where f is a form (homogeneous polynomial) of degree $a + b$ with coefficients in K.

It is not difficult to verify that the conditions (2.3.2a) of Proposition 2.3.2a are satisfied, and therefore the $n + 1$ affine spaces U_i may be glued together.

2.3.3b. Monoidal Transformation

With the notation of Sect. 2.3.3a., set

$$U_i = \operatorname{Spec} K\left[T_0, \ldots, T_n; \frac{T_0}{T_i}, \ldots, \frac{T_n}{T_i}\right],$$

$$U_{ij} = \operatorname{Spec} K\left[T_0, \ldots, T_n; \frac{T_0}{T_i}, \ldots, \frac{T_n}{T_i}\right]_{T_j/T_i}.$$

As above, the rings of functions on U_{ij} and U_{ji} can be identified with the ring whose elements are of the form $f(T_0, \ldots, T_n)/T_i^a T_j^b$, where f is now an inhomogeneous polynomial, in which the lowest degree terms have degree $\geq a + b$.

Denote by X the scheme obtained by gluing the U_i, and identify the U_i and U_{ij} with the corresponding open sets in X. We have

$$X = \bigcup_{0 \leq i \leq n} U_i, \quad U_{ij} = U_i \cap U_j.$$

Let us examine the structure of X in detail. The monomorphism

$$K[T_0, \ldots, T_n] \longrightarrow K\left[T_0, \ldots, T_n, \frac{T_0}{T_i}, \ldots, \frac{T_n}{T_i}\right]$$

determines the projection of the U_i to $\mathbb{A}_K^{n+1} = \operatorname{Spec} K[T_0, \ldots, T_n]$. Obviously, these projections are compatible on U_{ij}. In U_i, single out the open subset $D_i = D(T_i)$. Since

$$K[T_0, \ldots, T_n]_{T_i} = K\left[T_0, \ldots, T_n; \frac{T_0}{T_i}, \ldots, \frac{T_n}{T_i}\right],$$

it follows that D_i is isomorphically mapped onto the complement of the "coordinate hyperplane" $V(T_i)$ in $\mathbb{A}_K^{n+1}$. Therefore, X has an open subset $\bigcup_{0 \leq i \leq n} D_i$ isomorphic to $\mathbb{A}_K^{n+1} \setminus V(T_0, \ldots, T_n)$, and, if K is a field, this is just the complement of the origin in the $(n + 1)$-dimensional affine space $\mathbb{A}_K^{n+1}$.

What is the structure of $X \setminus \left(\bigcup_{0 \leq i \leq n} D_i\right)$? We have

$$X \setminus \bigcup_{0 \leq i \leq n} D_i = \bigcup_{0 \leq i \leq n} V(T_i), \quad \text{where} \quad V(T_i) \subset U_i.$$

Furthermore,

$$V(T_i) = \operatorname{Spec} K\left[T_0, \ldots, T_n, \frac{T_0}{T_i}, \ldots, \frac{T_n}{T_i}\right] / (T_i)$$

The ring $K\left[T_0, \ldots, T_n, \dfrac{T_0}{T_i}, \ldots, \dfrac{T_n}{T_i}\right]$ consists of the elements of the form

$$\frac{f(T_0, \ldots, T_n)}{T_i^a}, \text{ where } f \text{ is a polynomial in which}$$

the degrees of its lowest terms are $\geq a$.

Hence, the ideal (T_i) of this ring consists of elements of the same form, but with the degrees of the lowest terms of the numerator $\geq a + 1$. Therefore, it is easy to see that

$$K\left[T_0, \ldots, T_n, \frac{T_0}{T_i}, \ldots, \frac{T_n}{T_i}\right] / (T_i) \simeq K\left[\frac{T_0}{T_i}, \ldots, \frac{T_n}{T_i}\right],$$

and so

$$V(T_i) = \operatorname{Spec} K\left[\frac{T_0}{T_i}, \ldots, \frac{T_n}{T_i}\right].$$

The affine schemes $V(T_i)$ are glued together as in the preceding example; therefore, from the set-theoretical point of view, $X = \left(\bigcup_{0 \leq i \leq n} D_i\right) \cup \mathbb{P}_K^n$.

Thus, X is obtained from the $(n + 1)$-dimensional affine space

$$\mathbb{A}_K^{n+1} = \operatorname{Spec} K[T_0, \ldots, T_n]$$

by pasting the projective space $\mathbb{P}_K^n$ instead of the "origin" $V(T_0, \ldots, T_n)$.

2.3.3c Exercise

(1) Prove that $\Gamma(\mathbb{P}_K^n, \mathcal{O}_{\mathbb{P}_K^n}) = K$. Calculate $\Gamma(X, \mathcal{O}_X)$ for the scheme X constructed in Example 2.3.3b.
(2) Prove that if $\operatorname{Spec} K$ is irreducible, then so is $\mathbb{P}_K^n$.

2.3.4   Condition Necessary for Gluing Spectra

Let us give a simple necessary algebraic condition that must be fulfilled in order to be able to glue $\operatorname{Spec} A$ and $\operatorname{Spec} B$:

2.3.4a Proposition *Let A and B be rings without zero divisors. If there is an open subset U of $\operatorname{Spec} A$ such that $(U, \widetilde{A}|_U)$ is isomorphic to $(W, \widetilde{B}|_W)$, where W is an open subset of $\operatorname{Spec} B$, then the fields of fractions of A and B are isomorphic.*

If A and B are rings of finite type over a field or $\mathbb{Z}$, then the converse is also true.

Proof Consider an isomorphism $(U, \widetilde{A}|_U) \xrightarrow{\simeq} (W, \widetilde{B}|_W)$. Generic points of Spec A and Spec B are mapped into each other (they are contained in U, and therefore in W). The stalks of the structure sheaves at these points are the fields of fractions A and B, respectively.

To prove the converse statement, notice first of all that if A has no zero divisors, then Spec A and Spec $A\left[\dfrac{f}{g}\right]$, where $f, g \in A$, contain isomorphic big open sets:

$$A\left[\frac{1}{fg}\right] = A\left[\frac{1}{f}, \frac{1}{g}\right] = \left(A\left[\frac{f}{g}\right]\right)_{f/1}.$$

Now, if A (resp. B) is generated (over K or $\mathbb{Z}$) by elements $x_1, \ldots, x_n$ (resp. $y_1, \ldots, y_n$), and the fields of fractions of A and B are isomorphic, then we may pass from A (resp. B) to the ring $A[y_1, \ldots, y_n] = B[x_1, \ldots, x_n]$ by a finite number of steps, adjoining each time one element of the field of fractions and at each step the spectra of the considered rings have isomorphic open sets. Taking into account the irreducibility of Spec A and Spec B, we obtain the desired conclusion. $\square$

One of the immediate corollaries of this Proposition is the promised equivalence of the category **Aff Sch** of affine schemes, as defined in Sect. 1.5.2, with the category of affine schemes, as defined in Sect. 2.3.2—the full subcategory of the category of schemes Sch.

The schemes X and Y are said to be *birationally equivalent* if there exist everywhere dense open subsets $U \subset X$ and $V \subset Y$ such that $(U, \mathcal{O}_X|_U)$ is isomorphic to $(V, \mathcal{O}_X|_V)$.

The origin of the term "birational equivalence" is as follows. On the spectrum of any ring without zero divisors, the elements of its field of fractions can be considered as "rational functions". An isomorphism of open subsets of Spec A with open subsets of Spec B was interpreted as a "not everywhere defined" map given by rational functions.

2.3.4b Example Spec $k[T]$ (the line) and Spec $k[T_1, T_2]/(T_1^2 + T_2^2 - 1)$ (the circle) are birationally equivalent if Char $k \neq 2$. (**Exercise.** How can one show that they are not isomorphic?)

Indeed, the classical parametrization

$$t_1 = \frac{T^2 - 1}{T^2 + 1}, \quad t_2 = \frac{2T}{T^2 + 1}$$

and its inverse $T = \dfrac{t_2}{1 + t_1}$ establish an isomorphism of rings of fractions

$$k[T]_{T^2+1} \simeq (k[T_1, T_2]/(T_1^2 + T_2^2 - 1))_{1-t_1},$$

where $t_i = T_i \pmod{T_1^2 + T_2^2 - 1}$.

2.3.4c Example A generalization of the construction of the previous example. Let $f(T_1, \ldots, T_n)$ be an indecomposable quadratic polynomial over a field k of characteristic $\neq 2$, and with a zero over k. The spaces $X = \operatorname{Spec} k[T_1, \ldots, T_n]/(f)$ and $Y = \operatorname{Spec} k[T_1', \ldots, T_{n-1}']$ are birationally equivalent. We will describe a parametrization geometrically, leaving the details and the description of isomorphic open sets as an exercise for the reader.

Consider X as a subspace of the affine space $E = \operatorname{Spec} k[T_1, \ldots, T_n]$; let us embed Y into E by means of the ring homomorphism

$$T_i \longmapsto T_i' \text{ for } i = 1, \ldots, n-1, \text{ and } T_n \longmapsto 0.$$

On the geometric k-points of the spaces X and Y, the correspondence given by this parametrization is described as follows.

Let us fix a k-point x of the quadric X. We may assume that it does not lie on Y; otherwise we can modify the embedding of Y. Let us draw lines in E through the fixed point $x \in X$ and a variable point $y \in Y$. To every point $y \in Y$ we assign a distinct from x point of intersection z of the line $\overline{xy}$ with Y. The point z exists and is uniquely determined if y lies in a non-empty open subset of Y.

2.3.4d   Spec $k[T]$ and Spec $k[T_1, T_2]/(T_1^3 + T_2^3 - 1)$ Are Birationally Equivalent

To prove this, it suffices to establish that the equation $X^3 + Y^3 = Z^3$ has no solutions in $k[T]$, except those proportional to "constant" solutions (i.e., with $X, Y, Z \in k$). We may assume that the cubic roots of unity lie in k and apply the classical *Fermat's descent*[2] (with respect to the degree of the polynomial) using the unique factorization property of $k[T]$.

---

[2]From Wikipedia: Fermat's infinite descent is illustrated by the proof of Proposition 2.3.4d.i. It is essentially equivalent to Fermat's own proof, circa 1640, that the area of a Pythagorean triangle cannot be a square, which he described to Huygens simply by saying "if the area of such a triangle were a square, then there would also be a smaller one with the same property, and so on, which is impossible".

2.3.4d.i Proposition   *There are no integer solutions of $x^4 + y^4 = z^2$.*

Proof Suppose there are integers x, y, z such that $x^4 + y^4 = z^2$. This can be written as a Pythagorean triple $(x^2)^2 + (y^2)^2 = z^2$, from which it follows that $y^2 = 2pq$, $x^2 = p^2 - q^2$, and $z = p^2 + q^2$. Since $2pq$ is a square, we know that either p or q is even. Thus, from the Pythagorean triple $x^2 + q^2 = p^2$ we have $x = r^2 - s^2$, $q = 2rs$, and $p = r^2 + s^2$. Also, since $2pq$ is a square, we can set $q = 2u^2$ and $p = v^2$.

Now, since $2u^2 = 2rs$, we have $r = g^2$ and $s = h^2$. These, along with $p = v^2$, can be substituted back into $p = r^2 + s^2$ to give $v^2 = g^4 + h^4$, where $v < z$, contradicting the fact that there must be a smallest solution.   □

2.4 Projective Spectra

In this section, we introduce a very important class of schemes—the projective spectra of $\mathbb{Z}$-graded rings. This class contains analogues of classical projective varieties and, in particular, projective spaces.

2.4.1 $\mathbb{Z}$-graded Rings

First, recall the definition of a $\mathbb{Z}$-graded ring R (commutative and with a unit, as always in these lecture notes). Let $R = \bigoplus_{i \in \mathbb{Z}} R_i$ be a direct sum of commutative subgroups with respect to addition and let $R_i R_j \subset R_{i+j}$, in the sense that $r_i r_j \in R_{i+j}$ for any $r_i \in R_i$ and $r_j \in R_j$. The elements $r \in R_i$ are said to be *homogeneous of degree i*. The function $\deg \colon R \longrightarrow \mathbb{Z}$ is defined by the formula

$$\deg r = i \ \text{ if } \ r \in R_i.$$

Clearly, any nonzero $r \in R$ can be uniquely represented in the form $r = \sum r_i$ with $r_i \in R_i$.

An ideal $P \subset R$ is said to be *homogeneous* if $P = \bigoplus(P \cap R_i)$. For any homogeneous ideal P, the quotient ring is naturally $\mathbb{Z}$-graded: $R/P = \bigoplus(R_i/P_i)$. Clearly, R_0 is a subring and if $R_i = 0$ for $i < 0$, then $R_+ = \bigoplus_{i>0} R_i$ is a homogeneous ideal of R.

2.4.1a Example The *standard grading* of $R = K[T_0, \dots, T_n]$ is given by setting

$$\deg f = 0 \ \text{ for any } f \in K \text{ and } \deg T_i = 1 \text{ for all } i.$$

2.4.2 Projective Spectra

The *projective spectrum* of the $\mathbb{Z}$-graded ring R with $R_i = 0$ for $i < 0$ is the topological space

$$\operatorname{Proj} R = \{\text{homogeneous prime ideal } \mathfrak{p} \text{ of } R \mid \mathfrak{p} \not\supset R_+ = \bigoplus_{i>0} R_i\}$$

with the topology induced by the Zariski topology of $\operatorname{Spec} R$.

2.4.2a   A Geometric Interpretation

In classical projective geometry, a projective variety X over an algebraically closed field K is given by a finite system of homogeneous equations

$$F_i(T_0, \ldots, T_n) = 0, \quad \text{where } i \in I.$$

We associate with X the graded ring $R = K[T_0, \ldots, T_n]/(F_i)_{i \in I}$. Several schemes that can be constructed from $K[T_0, \ldots, T_n]$ are related with the following geometric objects (Fig. 2.4):

- $\mathbb{A}_K^{n+1} = \operatorname{Spec} K[T_0, \ldots, T_n]$, the $(n + 1)$-dimensional *affine space* over K with a fixed coordinate system.
- $C = \operatorname{Spec} R$, a subscheme of $\mathbb{A}_K^{n+1}$, a cone with the vertex at the origin. Indeed, the characteristic property of the cone is that together with every (geometric) point it contains the corresponding *generator*—the straight line through this point and the vertex of the cone. The straight line connecting the point $(t_0, \ldots, t_n)$ with the origin consists of the points $(tt_0, \ldots, tt_n)$ for any $t \in K$, and all these points belong to C, because C is given by homogeneous equations. Varying t we move along the generator. Moreover, every *non-zero* value of t determines an automorphism of R which multiplies the homogeneous elements of degree i by t^i. From this it is easy to deduce that, conversely, every cone is given by homogeneous equations.
- $\mathbb{P}_K^n = \operatorname{Proj} K[T_0, \ldots, T_n]$. By the above, to the points of $\mathbb{P}_K^n$ there correspond the irreducible cones in $\mathbb{A}_K^{n+1}$; in particular, to the closed points there correspond straight lines through the origin.

    This is the usual definition of the projective space. Though the origin in $\mathbb{A}_K^{n+1}$ is a homogeneous prime ideal, $R_+ = (T_0, \ldots, T_n)$, it only contains the vertex of the cone, and therefore is not regarded as a point of $\mathbb{P}_K^n$.

    It is convenient to assign to every straight line through the origin its point at infinity. Then, $\mathbb{P}_K^n$ can be interpreted as a *hyperplane at infinity in* $\mathbb{A}_K^{n+1}$.

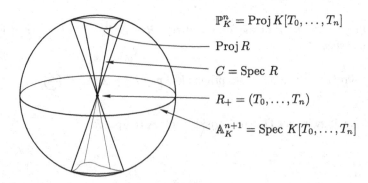

$$\mathbb{P}_K^n = \operatorname{Proj} K[T_0, \ldots, T_n]$$

$$\operatorname{Proj} R$$

$$C = \operatorname{Spec} R$$

$$R_+ = (T_0, \ldots, T_n)$$

$$\mathbb{A}_K^{n+1} = \operatorname{Spec} K[T_0, \ldots, T_n]$$

Fig. 2.4

- Proj R. The above discussion implies that Proj R corresponds to the base of the cone C, which belongs to the hyperplane at infinity, $\mathbb{P}^n_K$. □

2.4.3 The Scheme Structure on Proj R

Let us define the structure sheaf on Proj R and show that the obtained ringed space is locally affine.

For every $f \in R$, define big open sets as

$$D_+(f) := D(f) \cap \text{Proj } R.$$

Clearly, if $f = \sum_{i \in \mathbb{Z}_+} f_i$, where f_i are homogeneous elements of degree i, then $D_+(f) = \bigcup_{i \in \mathbb{Z}_+} D_+(f_i)$, and therefore the sets $D_+(f)$, where f runs over the homogeneous elements of R, form a basis of a topology on Proj R. For any homogeneous $f \in R$, the localization R_f, is, clearly, graded:

$$\deg(g/f^k) = \deg g - k \deg f. \tag{2.1}$$

2.4.3a Exercise Prove that the grading determined by relation (2.1) is well defined. (Use the isomorphism $R_f \simeq R[T]/(fT - 1)$ assuming that T is homogeneous of degree $-\deg f$.)

Denote by $(R_f)_0$ the component of degree 0. This component is of highest importance in the projective case, because, in contrast to the affine case, only the elements of $(R_f)_0$ are candidates for the role of "functions" on $D_+(f)$.

2.4.3b Proposition *Let f, g be homogeneous elements of R. Then*

(a) $D_+(f) \cap D_+(g) = D_+(fg)$;
(b) *there exists a collection of homeomorphisms $\psi_f : D_+(f) \longrightarrow \text{Spec}(R_f)_0$ such that all diagrams*

$$\begin{array}{ccc} D_+(f) & \xrightarrow{\psi_f} & \text{Spec}(R_f)_0 \\ \uparrow & & \uparrow \\ D_+(fg) & \xrightarrow{\psi_{fg}} & \text{Spec}(R_{fg})_0 \end{array}$$

commute.

(Here the left vertical arrow is the natural embedding, and the right one is induced by the natural ring homomorphism $R_f \longrightarrow R_{fg}$.)

2.4.3b.i Corollary *The sheaves $\psi_f^*((\widetilde{R_f})_0)$ transferred to $D_+(f)$ via ψ_f are glued together and determine a scheme structure on Proj R.*

Proof of Proposition 2.4.3b The fact that $D(f) \cap D(g) = D(fg)$ yields item (a).

To prove item (b), define ψ_f as the through map

$$D_+(f) \longrightarrow D(f) \longrightarrow \operatorname{Spec} R_f \longrightarrow \operatorname{Spec}(R_f)_0,$$

where the first arrow is the natural embedding, the second one is the isomorphism and the last one is induced by the ring monomorphism $(R_f)_0 \longrightarrow R_f$.

Clearly, ψ_f is continuous. Let us show that it is one-to-one: we construct the inverse map $\varphi \colon \operatorname{Spec}(R_f)_0 \longrightarrow D_+(f)$. Let $\mathfrak{p} \in \operatorname{Spec}(R_f)_0$. Set

$$\varphi(\mathfrak{p})_n = \{x \in R_n \mid x^d/f^n \in \mathfrak{p}\}, \text{ where } f \in R_d.$$

First, let us verify that $\varphi(\mathfrak{p}) = \bigoplus_n \varphi(\mathfrak{p})_n$ is a homogeneous prime ideal. Let $x, y \in \varphi(\mathfrak{p})_n$. Let us establish that $\varphi(\mathfrak{p})_n$ is closed with respect to addition; the other properties of the ideal are even easier to establish.

Clearly, if $x^d/f^n,\ y^d/f^n \in \mathfrak{p}$, then $(x+y)^{2d}/f^{2n} \in \mathfrak{p}$, and therefore $(x+y)^d/f^n \in \mathfrak{p}$, so $x + y \in \varphi(\mathfrak{p})$. We have taken into account that $\mathfrak{p}$ is a prime ideal in $(R_f)_0$. Furthermore, if $x^d/f^n,\ y^d/f^m \in \varphi(\mathfrak{p})_{m+n}$, then $(xy)^d \in \mathfrak{p}$, implying that either $x \in \mathfrak{p}$ or $y \in \mathfrak{p}$, because $\mathfrak{p}$ is prime; therefore either x^d/f^n or y^d/f^m belongs to $\varphi(\mathfrak{p})$. Hence $\varphi(\mathfrak{p})$ is prime.

2.4.3c Exercise Verify that φ and ψ_f are mutually inverse.

Now, let us prove that ψ_f is a homeomorphism. It suffices to show that ψ_f is an open map, because its continuity is already established. Let $g \in R_f$. We have to verify that the image of $D_+(f) \cap D_+(g)$ under ψ_f is open:

$$D_+(f) \cap D_+(g) = D_+(fg) \longrightarrow \operatorname{Spec}(R_{fg})_0 =$$
$$= \operatorname{Spec}((R_f)_0)_{(g^d/f^e)} \longrightarrow \operatorname{Spec}(R_f)_0,$$

which also shows that one can glue $\operatorname{Spec}(R_f)_0$ and $\operatorname{Spec}(R_g)_0$ along $D(fg)$, because

$$\operatorname{Spec}(R_f)_0 \supset D_+(fg) = \operatorname{Spec}(R_{fg})_0 = \operatorname{Spec}((R_f)_0)_{(g^d/f^e)}$$
$$= \operatorname{Spec}((R_g)_0)_{(f^e/g^d)} = D_+(gf) \subset \operatorname{Spec}(R_g)_0. \qquad \square$$

2.4.4 *Examples*

(1) $\operatorname{Proj} K[x_0, \ldots, x_n]$ is the projective space $\mathbb{P}^n_K$ constructed in Sect. 2.3.3a.
(2) The scheme X from Example 2.3.3b. can be alternatively obtained as follows. In $K[T_0, \ldots, T_n]$, consider the ideal $J = (T_0, \ldots, T_n)$. Denote by $R_k = J^k T^k$ the set of monomials of degree k in T with coefficients in the kth power of J; set $R = \bigoplus R_k$. Clearly, $X = \operatorname{Proj} R$.

A generalization of this construction goes as follows.

(3) **The monoidal transformation with center in an ideal**. Let R be a ring and J its ideal. In $R[T]$, construct the graded subring:

$$R = \bigoplus_{k\geq 0} R_k, \quad \text{where } R_k = J^k T^k,$$

i.e., the elements of R are the polynomials $\sum a_k T^k$ such that $a_k \in J^k$.

We say that $\operatorname{Proj} R$ is the result of applying to $\operatorname{Spec} R$ a *monoidal transformation with center in J*.   □

2.4.5 Two Essential Distinctions Between Affine and Projective Spectra

(1) Not every homomorphism of graded rings $f\colon R \longrightarrow R'$, not even a homogeneous one, necessarily induces a map $\operatorname{Proj} R' \longrightarrow \operatorname{Proj} R$. For example, consider the monomorphism

$$K[T_0, T_1] \longrightarrow K[T_0, T_1, T_2], \qquad T_i \longmapsto T_i.$$

Then, the ideal $(T_0, T_1) \subset K[T_0, T_1, T_2]$ has no preimage in $\operatorname{Proj} \ K[T_0, T_1]$.

The corresponding geometric picture looks like this: the projection $(t_0 : t_1 : t_2) \longmapsto (t_0 : t_1)$ of the plane onto the straight line is not defined at $(0 : 0 : 1)$ because the point $(0 : 0)$ does not exist on the projective line.

(2) We have established a one-to-one correspondence between rings and affine schemes: from A we recover the scheme $(\operatorname{Spec} A, \widetilde{A})$, and from an affine scheme $(X, \mathcal{O}_X)$ we recover the ring of global functions on it: $A = \Gamma(X, \mathcal{O}_X)$.

However, the analogue of this statement fails for projective spectra: it may very well happen that $\operatorname{Proj} R_1 \simeq \operatorname{Proj} R_2$ for quite different rings R_1 and R_2. (Examples of such phenomena are given in the next subsection.)

2.4.6 Properties of R Which Reflect Certain Properties of Proj R

The rather extravagant, from the algebraist's point of view, relation

$$R_1 \sim R_2 \Longleftrightarrow \operatorname{Proj} R_1 \cong \operatorname{Proj} R_2$$

adds a geometric flavor to our Algebra.

Here are two ways to vary R while preserving $X = \operatorname{Proj} R$:

(1) For any $\mathbb{Z}$-graded ring R, define its *dth Veronese ring* $R^{(d)}$, where $d \in \mathbb{N}$, by setting $(R^{(d)})_i = R_{di}$.
(2) Take $R \subset R'$ such that $R_i = R'_i$ for all $i \geq i_0$.

2.4.6.i Lemma

(1) $\operatorname{Proj} R' \cong \operatorname{Proj} R$.
(2) $\operatorname{Proj} R^{(d)} \cong \operatorname{Proj} R$.

Proof
(a) Indeed, $\operatorname{Proj} R = \bigcup_{\deg f \geq i_0} D_+(f) = \bigcup_{\deg f \geq i_0} \operatorname{Spec}(R_f)_0$; moreover, any element of $(R_f)_0$ can be represented in the form g/f^n with $\deg g \geq i_0$ (multiply both g and f^n by a sufficiently high power of f).
(b) Define the homeomorphism $\operatorname{Proj} R \cong \operatorname{Proj} R^{(d)}$ by setting $\mathfrak{p} \longmapsto \mathfrak{p} \cap R^{(d)}$. For the sheaves, see the argument in the proof of (a).   □

2.5   Algebraic Invariants of Graded Rings

Unless otherwise stated, we will consider only the following simplest case of $\mathbb{Z}$-graded rings R:

(1) $R_i = 0$ for $i < 0$ and $R_0 = K$ is a field;
(2) R_1 is a finite-dimensional space over K;
(3) R_1 generates the K-algebra R.

The schemes $\operatorname{Proj} R$ for such rings R are the closest to the classical notion of a projective algebraic variety over K. In particular, if $\dim_K R_1 = r + 1$, then the epimorphism of $\mathbb{Z}$-graded rings $K[T_0, \dots, T_r] \longrightarrow R$ sending $T_0, \dots, T_r$ to a K-basis of R_1 determines an embedding

$$\operatorname{Proj} R \longhookrightarrow \operatorname{Proj} K[T_0, \dots, T_r] = \mathbb{P}^r_K.$$

We will introduce certain invariants of R following the simple and beautiful idea of Hilbert to study $h_R(n) := \dim_K R_n$ as a function of n.

2.5.1 $\dim_K R_n$ is a Polynomial with Rational Coefficients for Some $n \geq n_0 = n_0(R)$

We will prove a more general statement making use of the following notions. An R-module M is said to be *$\mathbb{Z}$-graded* if $M = \bigoplus_{i \in \mathbb{Z}} M_i$ and $R_i M_j \subset M_{i+j}$.

A homomorphism $f: M \longrightarrow N$ of $\mathbb{Z}$-graded modules is called *homogeneous of degree d* if $f(M_i) \subset N_{i+d}$.

2.5.2 Theorem *Let M be a $\mathbb{Z}$-graded R-module with finitely many generators. Then $\dim_K M_n = h_M(n)$ for $n \geq n_0 = n_0(M)$, where $h_M(n)$ is a polynomial with rational coefficients.*

Proof Induction on $\dim_K R_1 = r$.

For $r = 0$, we have $R = K$ and M is a usual finite-dimensional linear space. Clearly, in this case, for a sufficiently large n (greater than the maximal degree of the generators of M), we have $\dim_K M_n = 0$, and the desired polynomial is zero.

The induction step: let the statement hold for $\dim_K R_1 \leq r-1$. Let x be a nonzero element of R_1. Then the (say, left) action $l_x: M \longrightarrow M$ of x on M is a homomorphism of degree 1. Consider the following exact sequence, where K_n and C_n are the degree-n homogeneous components of the kernel and cokernel of l_x, respectively:

$$0 \longrightarrow K_n \longrightarrow M_n \xrightarrow{\ l_x\ } M_{n+1} \longrightarrow C_{n+1} \longrightarrow 0$$

Clearly, $K = \mathrm{Ker}\, l_x = \bigoplus K_n$ and $C = \mathrm{Coker}\, l_x = \bigoplus C_n$ are $\mathbb{Z}$-graded $R/(x)$-modules. We have

$$\dim(R/(x))_1 = \dim R_1/Kx = \dim R_1 - 1.$$

By the induction hypothesis

$$h_M(n + 1) - h_M(n) = h_C(n + 1) - h_K(n) = h'(n). \tag{2.2}$$

Adding up the identities (2.2) starting from some $n = n_0$ we get the result desired if we take into account the following elementary fact: *the sum $\sum_{n=n_0}^{N} n^i$, considered as a function of the upper limit of summation, is a degree $i+1$ polynomial in N.* $\square$

The polynomial $h_M(n)$ is called the *Hilbert polynomial* of the R-module M. In particular, for $M = R$, we get the *Hilbert polynomial* of R.

The number $\deg h_R(n)$ is called the *dimension* of the projective spectrum $X = \mathrm{Proj}\, R$. (We do not claim yet that it depends only on X!)

2.5.3 Lemma *Let $h(x) \in \mathbb{Q}[x]$ take integer values for integer x. Then*

$$h(x) = \sum_{i \geq 0} a_i \frac{x(x-1)\cdots(x-i+1)}{i!}, \quad \text{where } a_i \in \mathbb{Z}.$$

Proof Every polynomial from $\mathbb{Q}[x]$ can be represented in the form

$$h(x) = \sum_{i \geq 0} a_i \frac{x(x-1)\cdots(x-i+1)}{i!},$$

where $a_i \in \mathbb{Q}$ and $a_0 = h(0) \in \mathbb{Z}$. By induction on i we get

$$h(i) = a_i + \sum_{j \leq i-1} a_j \frac{i \cdots (i-j+1)}{i!},$$

implying $a_i \in \mathbb{Z}$.   □

The leading coefficient of the polynomial

$$h(x) = \sum_{i \geq 0} a_i \frac{x \cdots (x-i+1)}{i!}$$

is called the *degree of the projective scheme* $\operatorname{Proj} R$.

The constant term $h_R(0)$ is called the *(Euler) characteristic* of $X = \operatorname{Proj} R$ and is denoted by $\chi(X)$.

The *arithmetic genus* of X is $p_a(X) = (-1)^{\dim X}(\chi(X) - 1)$.

2.5.3a Exercise $\dim \mathbb{P}_K^r = r$, $\deg \mathbb{P}_K^r = 1$, $p_a(\mathbb{P}_K^r) = 0$.

2.5.3b Remark Actually, **the whole** Hilbert polynomial is a (projective) invariant of the projective spectrum X. We have selected just the degree of h_X and two of its coefficients because $\dim X$ and $\chi(X)$ actually depend on X, not R, as we will show in what follows. The degree of h_X is the most important projective invariant which, together with the dimension, participates in the formulation of the Bézout theorem, see Sect. 2.5.6a. The geometric meaning of other coefficients is not known, except for the following result.

R. Hartshorne proved a beautiful theorem according to which the Hilbert polynomial is the only "discrete" projective invariant in the following sense:

Two projective schemes X and Y have the same Hilbert polynomials if and only if they can be "algebraically deformed" into each other inside some projective space.

2.5.4 Properties of the Degree

Let us make the convention that the degree of the zero polynomial is equal to 1.

2.5.4a Lemma *The following statements are equivalent:*

(1) $\dim X = -1$;
(2) $X = \emptyset$;
(3) $R_1 \subset \mathfrak{n}(R)$, *where* $\mathfrak{n}(R)$ *is the nilradical of* R.

Proof (1) $\Longrightarrow$ (3) is obvious because in this case $R_1^n = 0$ for $n \geq n_0$.

(3) $\Longrightarrow$ (1) follows from the fact that $\dim_K R_1 < \infty$ and R is generated by R_1.

(3) $\implies$ (2) follows from the equalities

$$X = \bigcup_{f \in R_+} D_+(f) = \bigcup_{f \in R_+} \mathrm{Spec}(R_f)_0$$

and the fact that $\mathrm{Spec}(R_f)_0 = \emptyset$, because the localization with respect to any multiplicative set containing a nilpotent element is the zero ring 0.

(2) $\implies$ (3) is obvious because in this case $\mathrm{Spec}(R_f)_0 = \emptyset$ for any $f \in R_+$ which means exactly that $(R_f)_0 = 0$, i.e., $f^n = 0$ for some n, for all f. □

2.5.4b Lemma *Let R be an arbitrary $\mathbb{Z}$-graded ring, and $\{f_i\}_{i \in I}$ a collection of its elements. Then the following statements are equivalent:*

(1) $X = \bigcup_i D_+(f_i)$;
(2) *for any $g \in R_+$ and some n, we have $g^n \in \sum_i Rf_i$.*

Proof Let $X = \bigcup_i D_+(f_i)$ and let $g \in R_+$ be a homogeneous element. Then $D_+(g) = \bigcup_i D_+(f_i g)$ by the definition of D_+. Further,

$$\mathrm{Spec}(R_g)_0 = \bigcup_i \mathrm{Spec}((R_g)_0)_{f_i^d/g^l}$$

implying

$$1/1 = \sum (f_i^d/g^l) a_i, \text{ where } a_i \in (R_g)_0,$$
$$g^n = \sum f_i^{m_i} b_i, \qquad \text{where } b_i \in R.$$

Since all these arguments are reversible, (1) and (2) are equivalent. □

2.5.4b.i Corollary *If R is generated by R_1 as an R_0-algebra, then for any set of generators $\{f_i\}_{i \in I}$ of the R_0-module R_1, we have $\mathrm{Proj}\, R = \bigcup_i D_+(f_i)$.*

2.5.4c Proposition *Let R satisfy the condition formulated at the beginning of this section.*

(1) *The following conditions are equivalent:*

$$\dim \mathrm{Proj}\, R = 0 \iff \mathrm{Proj}\, R \text{ is finite.}$$

(2) *If these conditions hold, then the topology on $\mathrm{Proj}\, R = X$ is discrete, and, as schemes,*

$$\mathrm{Proj}\, R \cong \mathrm{Spec}\, \Gamma(X, \mathcal{O}_X),$$

where $\Gamma(X, \mathcal{O}_X)$ is a K-algebra of finite rank. In this case

$$\chi(X) := \deg X = \dim_K \Gamma(X, \mathcal{O}_X).$$

Proof
(1) Let us show that if $\dim X = 0$, then X is finite and there is an n_0 such that

$$\dim_K R_n = \dim \Gamma(X, \mathcal{O}_X) \text{ for any } n \geq n_0. \tag{2.3}$$

Indeed, if $\dim X = 0$, then $\dim R_n = d \neq 0$ for $n \geq n_0$, implying that $\dim_K R_{(f)} < \infty$ for every $f \in R_1$. Otherwise, there would have existed an element $g \in R_1$ such that $g/f, (g/f)^2, \ldots, (g/f)^n$ were linearly independent over K; but this is impossible, because then the elements $g^i f^{n-i}$ for $1 \leq i \leq n$ would be linearly independent in R_n for arbitrarily large n, in particular, for $n > d$.

Since $\dim_K (R_f)_0 < \infty$, we deduce that $\mathrm{Spec}(R_f)_0$ is finite and discrete. Indeed, any prime ideal in $(R_f)_0$ is maximal because the quotient by this ideal is a finite-dimensional algebra over K without zero divisors, i.e., a field. Therefore, any prime ideal of $(R_f)_0$ is minimal, and since $(R_f)_0$ is clearly Noetherian, there are only finitely many minimal ideals.

The space $\mathrm{Proj}\, R$ can be covered by finitely many discrete opens $\mathrm{Spec}(R_f)_0$, where f runs over a K-basis of R_1, and therefore $\mathrm{Proj}\, R$ is finite and discrete. It follows that $\Gamma(X, \mathcal{O}_X) = \prod_{x \in X} \mathcal{O}_x$ and $\dim_K \mathcal{O}_x < \infty$. This immediately implies an isomorphism of ringed spaces

$$X \simeq \mathrm{Spec}\, \Gamma(X, \mathcal{O}_X).$$

Observe that there exists a homogeneous element $f \in R_+$ such that $D_+(f) = X$. Indeed, assume the contrary. Then every element of R_+ vanishes at one of the points of X. Select a minimal subset of points $Y \subset X$ such that every element of R_+ vanishes at one of the points of Y. Since the ideals, which are points of $\mathrm{Proj}\, R$, do not contain R_+, it follows that Y contains more than one point. For every $y \in Y$, there exists an element $a_y \in R_+$ that vanishes nowhere on Y, except at y. Then $b_x = \prod_{y \in Y \setminus \{x\}} a_y$ vanishes everywhere on Y except at x, whereas $\sum_{x \in X} b_x$ vanishes nowhere on Y: a contradiction.

Let $D_+(f) = X$; then $\Gamma(X, \mathcal{O}_X) = R_{(f)}$ and it suffices to establish that

$$\dim_K R_{(f)} = \dim_K R_n \text{ for all } n \geq n_0.$$

We have $(R_f)_0 = \bigcup R_n/f^n$, and $R_n/f^n \subset R_{n+1}/f^{n+1}$. Since $\dim_K(R_f)_0$ is finite, we see that $(R_f)_0 = R_n/f^n$ for all $n \geq n_0$, and therefore

$$\dim_K(R_f)_0 = \dim_K R_n.$$

To prove equality (2.3), it suffices to verify that $\dim_K(\bigcup_m \mathrm{Ann}\, f^m) < \infty$; then, for n large, the map $g \longmapsto g/f^n$, where $g \in R_n$, is an isomorphism.

Indeed, since $X = D_+(f)$, then for any $g \in R_+$, there exists a k such that $g^k \in fR$. Since $\dim_K R < \infty$, we can choose k independent of g, and therefore $R_n = fR_{n-1}$ for sufficiently large n. As $\dim_K R_n = \dim_K R_{n-1}$, no power of f is annihilated by elements of sufficiently high degree.

(2) It remains to verify that X is infinite if $\dim X > 0$. Indeed, if $\dim X > 0$, then $\dim_K R_n \longrightarrow \infty$ as $n \longrightarrow \infty$, implying that $\dim_K (R_f)_0 = \infty$ for some $f \in R_1$: otherwise the same arguments as above lead to a contradiction.

Therefore, it suffices to show that $\operatorname{Spec} A$ is infinite for any K-algebra A with finitely many generators such that $\dim_K A = \infty$. By Noether's normalization theorem 1.5.3f, A is a finitely generated module over a subalgebra of it isomorphic to $K[T_1, \ldots, T_d]$; because $\dim_K A = \infty$, we have $d > 0$. Now, Theorem 1.6.5 and the fact that $\operatorname{card}(\operatorname{Spec} K[T_1, \ldots, T_d]) = \infty$ imply that $\operatorname{card}(\operatorname{Spec} A) = \infty$.   □

2.5.5 Characteristic Functions and Bézout's Theorem

Let R be a ring satisfying the conditions of the previous section, $h_R(n)$ its Hilbert polynomial.

2.5.5a Proposition

(1) *The generating function*

$$F_R(t) = \sum_{n \geq 0} h_R(n) t^n$$

is a rational function in t; moreover, the degree of the pole of $F_R(t)$ at $t = 1$ is equal to $1 + \dim \operatorname{Proj} R$;

(2) $\chi(\operatorname{Proj} R) = -\operatorname*{Res}_{t=1} \dfrac{F_R(t)}{t} dt$;

(3) $\deg \operatorname{Proj} R = \lim\limits_{t \to 1} (t-1)^{\dim \operatorname{Proj} R + 1} F_R(t)$.

Proof For $k \geq 1$, we have

$$\sum_{n \in \mathbb{Z}_+} n^k t^n = \left(t \frac{d}{dt} \right) \sum_{n \in \mathbb{Z}_+} n^{k-1} t^n = \cdots = \left(t \frac{d}{dt} \right)^k \frac{1}{1-t},$$

implying

$$\sum_{n \in \mathbb{Z}_+} h_R(n) t^n = h_R \left(t \frac{d}{dt} \right) \frac{1}{1-t}.$$

Induction on k shows that

$$\left(t\frac{d}{dt}\right)^k \left(\frac{1}{1-t}\right) = \frac{k!}{(1-t)^{k+1}} + \cdots,$$

where the dots stand for the terms with poles of orders $\geq k$ at $t = 1$. This and the definitions of deg and χ imply (1) and (3).

To prove (2), observe that

$$\operatorname*{Res}_{t=1}\left(-\frac{h_R(0)}{t(1-t)}\,dt\right) = h_R(0) = \chi(\operatorname{Proj} R),$$

and for $h_R(n) = n^k$, where $k > 0$, we have

$$F_R(t) = \left(t\frac{d}{dt}\right)^k \frac{1}{1-t},$$

implying

$$\operatorname*{Res}_{t=1}\left(\frac{F_R(t)}{t}\,dt\right) = \operatorname*{Res}_{t=1} d\left(\left(t\frac{d}{dt}\right)^{k-1}\frac{1}{1-t}\right) = 0. \qquad \square$$

2.5.6 Example Suppose $f \in R_d$ is not a zero divisor. Given $F_R(t)$, it is easy to calculate $F_{R/fR}(t)$:

$$h_{R/fR}(n) = h_R(n) - h_R(n-d);$$

hence

$$F_{R/fR}(t) = \sum_{n=0}^{\infty}(h_R(n) - h_R(n-d))t^n = (1 - t^d)F_R(t) + P(t),$$

where $P(t)$ is a polynomial. (Without assuming that f is not a zero divisor we only get a coefficient-wise inequality $F_{R/fR}(t) \geq (1 - t^d)F_R(t) + P(t)$.) In particular,

$$\dim\operatorname{Proj}(R/fR) = \dim\operatorname{Proj} R - 1,$$
$$\chi(\operatorname{Proj}(R/fR)) = \chi(\operatorname{Proj} R) - h_R(-d),$$
$$\deg\operatorname{Proj}(R/fR) = d\deg\operatorname{Proj} R.$$

The scheme $Y = \operatorname{Proj}(R/fR)$ can be naturally embedded into $\operatorname{Proj} R$ as $V_+(f)$; since it is given by a single equation, it is called a *hypersurface* in $\operatorname{Proj} R$.

A particular case: let $R = K[T_0, \ldots, T_r]$, where the $T_i \in R_1$, and let $\mathbb{P}^r = \operatorname{Proj} R$. Induction on r gives

$$F_{\mathbb{P}^r}(t) = \frac{1}{1-t} F_{\mathbb{P}^{r-1}}(t) = \cdots = \frac{1}{(1-t)^{r+1}},$$

whence

$$h_{\mathbb{P}^r}(n) = \frac{1}{n!} \frac{d^n}{dt^n} \frac{1}{(1-t)^{r+1}} = \binom{n+r}{r},$$

$$\dim \mathbb{P}^r = r, \quad \deg \mathbb{P}^r = 1, \quad \chi(\mathbb{P}^r) = 1.$$

2.5.6a Theorem (Bézout's Theorem) *Let* $f_1, \ldots, f_s \in R = K[T]$, *where* $T = (T_0, \ldots, T_r)$, *be homogeneous polynomials of degrees* $d_1, \ldots, d_s$, *respectively. Let* $Y = \operatorname{Proj} R/(f_1, \ldots, f_s)$. *Then*

$$\begin{aligned} &\dim Y \geq r - s \ \text{ and } \ \deg Y \geq d_1 \cdots d_s, \\ &\text{where } \deg Y = d_1 \cdots d_s \text{ if } \dim Y = r - s. \end{aligned} \tag{2.4}$$

If f_{i+1} *is not a zero divisor in* $R/(f_1, \ldots, f_i)$ *for all* $i = 1, \ldots s-1$ *(recall that such* Y *is called a complete intersection, see Sect. 1.11.2), the inequalities (2.4) turn into equalities.*

In particular, if Y *is a zero-dimensional complete intersection, i.e.,* $r = s$, *then* $\deg Y = d_1 \cdots d_s$.

Proof Induction on s.   □

2.5.6b A Geometric Interpretation of Complete Intersections

Since, set-theoretically, one has $Y = \operatorname{Proj} R/(f_1, \ldots, f_s) = \bigcap_{1 \leq i \leq s} V_+(f_i)$, one should visualize Y as the intersection of the hypersurfaces singled out by the equations $f_i = 0$ in $\mathbb{P}^r_K$.

The condition "f_{i+1} is not a zero divisor in $R/(f_1, \ldots, f_i)$" geometrically means that the $(i + 1)$th hypersurface is in "general position" with the intersection of the preceding i hypersurfaces, i.e., it does not entirely contain any of the components of this intersection, cf. the geometric interpretation of zero divisors in Theorem 1.7.10a.

When $\dim Y = 0$, the formula

$$\deg Y = \dim_K \Gamma(Y, \mathcal{O}_Y) = \sum_{y \in Y} \dim_K \mathcal{O}_y$$

replaces the notion of "the number of intersection points, counting multiplicities".

If the field K is algebraically closed, the multiplicity of $y \in Y$ is by definition equal to the rank of the local ring $\mathcal{O}_y$.

The term "complete intersection" is connected with the following images. In $\mathbb{P}^3$ (over $\mathbb{R}$ or $\mathbb{C}$), let f_1 determine a non-degenerate quadric, and f_2 determine its tangent plane at some point x. The intersection of the surfaces $f_1 = 0$ and $f_2 = 0$ consists of two straight lines through x (the generators of the hyperboloid). These two straight lines constitute a complete intersection of the quadric and the plane. If we wish to single out one of them, say L, we have to take for f_3, for example, the equation of the plane through the line, and it is not difficult to see that f_3 is a zero divisor in $R/(f_1, f_2)$. This straight line L is not a complete intersection inside of the quadric because a complete intersection should be of degree ≥ 2 whereas the degree of the straight line is 1.

2.5.6c A Geometric Complete Intersection

There is an interesting version of the notion of complete intersection. Let, for definiteness, $R = K[T_0, \ldots, T_r]$ and $P \subset R$ a homogeneous prime ideal. The scheme $X = \mathrm{Proj}(R/P)$ is called a *geometric complete intersection* if there exists an ideal $P' \subset P$ such that $X' = \mathrm{Proj}(R/P')$ is a complete intersection and $\mathfrak{r}(P') = P$. The last condition means that the space of X' is the same as that of X, the only difference being the presence of nilpotents in the structure sheaf of X'.

2.5.6d Problem Is every scheme of the form $X = \mathrm{Proj}\, R/P$ a geometric complete intersection? The answer is not known even for curves in the three-dimensional space: can one define any irreducible curve by two equations?[3]

If $\dim X = r - 1$, where $X \subset \mathbb{P}^r$, the answer is positive:

2.5.6e Proposition *If $X \subset \mathbb{P}^r$ and the dimension of every irreducible component of X is $r - 1$, then X is a geometric complete intersection, i.e., X is given by a single equation.*

Proof It suffices to consider the case where X is irreducible.

Let $X = \mathrm{Proj}\, K[T_0, \ldots, T_r]/P$, where P is a prime ideal, and $f \in P$ an irreducible homogeneous element (obviously, such an f always exists). Then

$$R = K[T_0, \ldots, T_r]/(f), \quad \text{i.e., } P = fK[T_0, \ldots, T_r].$$

Indeed, there exists a natural epimorphism

$$K[T_0, \ldots, T_r]/(f) \longrightarrow K[T_0, \ldots, T_r]/P.$$

If it had a nontrivial kernel, then, because f is irreducible, we would have had

$$\dim K[T_0, \ldots, T_r]/P < \dim K[T_0, \ldots, T_r]/(f),$$

contradicting the fact that $\dim X = r - 1$. $\square$

---

[3]To define the curve by three equations is relatively easy, see [Sh0, Ex. 9 in §6, v.1].

Proposition 2.5.6e yields, in particular, the following description of points in $\mathbb{P}_K^r$. They are of three types:

(a) The generic point of $\mathbb{P}_K^r$ corresponding to the zero ideal of $K[T_0, T_1, T_2]$.
(b) The generic points of irreducible "curves", one-dimensional irreducible sets. They are in one-to-one correspondence with the irreducible forms in three indeterminates (up to a nonzero constant factor).
(c) Closed points. As follows from Hilbert's Nullstellensatz, if K is algebraically closed, such points are in one-to-one correspondence with the nonzero triples $(t_1:t_2:t_3)$ of elements of K determined up to a nonzero constant factor.

Let f_1, f_2 be two forms; then f_2 is not a zero divisor in $R/f_1 R$ if and only if f_1 and f_2 are coprime, i.e., if and only if the curves $f_1 = 0$ and $f_2 = 0$ have no common irreducible components.

Comparing the results of Sects. 2.5.4c and 2.5.6a, we get the classical Bézout's theorem for $\mathbb{P}^2$:

2.5.7 Theorem *If two curves on $\mathbb{P}^2$ have no common irreducible components, then the number of their intersection points* (counting multiplicities) *is equal to the product of their degrees.*

2.6 Presheaves and Sheaves of Modules

The sheaves of modules over schemes arise naturally in algebraic geometry as a generalization of the notion "module over a commutative ring"; a more careful analysis of this correspondence leads to distinguishing the so-called *quasi-coherent* sheaves of modules. As we will show, the quasi-coherent sheaves over Spec A are indeed in a one-to-one correspondence with the A-modules.

From the geometric point of view, the sheaves of modules over a ringed space X embody the intuitive notion of a continuous family of linear spaces parameterized by X. If the sheaf is isomorphic to the direct sum of finitely many copies of $\mathcal{O}_X$, then this family is "constant": the total space of the corresponding bundle is the direct product of X by the fiber. If the sheaf is locally isomorphic to such a direct sum, then it corresponds to a locally trivial vector bundle. In the general case even the dimensions of the fibers may vary.

2.6.1 Presheaves and Sheaves of Modules over a Ringed Space $(X, \mathcal{O}_X)$

Here we describe some main notions and results of sheaf theory which do not depend on the assumption that X is a scheme; because of their generality, they are not deep.

A *presheaf of modules over* $(X, \mathcal{O}_X)$, or a *presheaf of $\mathcal{O}_X$-modules*, is a presheaf $\mathcal{P}$ of Abelian groups such that each group $\Gamma(U, \mathcal{P})$ carries a $\Gamma(U, \mathcal{O}_X)$-module structure and these structures are compatible with the restriction maps:

$$r_V^U(sp) = r_V^U(s)r_V^U(p) \quad \text{for any} \quad U \supset V, \; p \in \mathcal{P}(V), \; s \in \mathcal{O}_X(V).$$

Let $\mathcal{P}_1$ and $\mathcal{P}_2$ be two presheaves of modules over $(X, \mathcal{O}_X)$. A *morphism of presheaves of modules* $f \colon \mathcal{P}_1 \longrightarrow \mathcal{P}_2$ is a set of $\Gamma(U, \mathcal{O}_X)$-module homomorphisms $f(U) \colon \mathcal{P}_1(U) \longrightarrow \mathcal{P}_2(U)$ given for any open set $U \subset X$ and commuting with the restriction maps.

It is not difficult to verify that *the presheaves of modules over $(X, \mathcal{O}_X)$ constitute an Abelian category.* In particular, for any presheaf morphism $f \colon \mathcal{P}_1 \longrightarrow \mathcal{P}_2$, the presheaves $\mathcal{K}erf$ and $\mathcal{C}okerf$ exist and are described by their groups of sections calculated locally:

$$(\mathcal{K}erf)(U) := \mathcal{K}erf(U),$$
$$(\mathcal{C}okerf)(U) := \mathcal{C}okerf(U)$$

(with obviously defined restriction maps).

Given two presheaves of $\mathcal{O}_X$-modules $\mathcal{P}_1$ and $\mathcal{P}_2$, define their *tensor product* $\mathcal{P}_1 \otimes_{\mathcal{O}_X} \mathcal{P}_2$ by setting

$$(\mathcal{P}_1 \otimes_{\mathcal{O}_X} \mathcal{P}_2)(U) := \mathcal{P}_1(U) \otimes_{\mathcal{O}_X(U)} \mathcal{P}_2(U)$$

(with obviously defined restriction maps).

Given a set I, define the *direct sum* (*direct product* if $|I| \geq |\mathbb{Z}|$) of $|I|$ copies of a presheaf $\mathcal{P}$ by setting

$$\mathcal{P}^{(I)}(U) = \prod_{i \in I} \mathcal{P}_i(U), \quad \text{where } \mathcal{P}_i \text{ is the } i\text{th copy of } \mathcal{P}.$$

The *direct sum* (product) of distinct presheaves is obviously defined. If $|I| = n \in \mathbb{N}$, we write $\mathcal{O}_X^{(n)}$ instead of $\mathcal{O}_X^{(I)}$.

A presheaf of $\mathcal{O}_X$-modules $\mathcal{P}$ is called a *sheaf of $\mathcal{O}_X$-modules* if $\mathcal{P}$ satisfies the axioms of a sheaf, i.e., it is a sheaf. If $\mathcal{P}$ is a presheaf of $\mathcal{O}_X$-modules and $\mathcal{P}^+$ the associated sheaf, then $\mathcal{P}^+$ also acquires the natural structure of a sheaf of $\mathcal{O}_X$-modules: originally, the sets $\mathcal{P}^+(U)$ are only defined as Abelian groups; however, the multiplication by the sections of $\mathcal{O}_X$ commutes with both limit procedures in the definition of the sheaf $\mathcal{P}^+$.

There is defined a canonical morphism of presheaves of modules $\mathcal{P} \longrightarrow \mathcal{P}^+$ because the image of $\mathcal{P}(U)$ under the homomorphism $\mathcal{P}(U) \longrightarrow \prod_{x \in U} \mathcal{P}_x$ belongs to $\mathcal{P}^+(U)$.

Every sheaf of $\mathcal{O}_X$-modules $\mathcal{F}$ can be regarded as a presheaf; the presheaf obtained in this way from a sheaf $\mathcal{F}$ will be denoted by $i(\mathcal{F})$. Defining a sheaf morphism $\mathcal{F}_1 \longrightarrow \mathcal{F}_2$ as a morphism of corresponding presheaves, we can consider i

as a functor which embeds the category of sheaves of $\mathcal{O}_X$-modules into the category of presheaves. This "tautological" functor is associated with a far less trivial functor $\mathcal{P} \longmapsto \mathcal{P}^+$ acting in the opposite direction as follows: for any presheaf $\mathcal{P}$ and any sheaf $\mathcal{F}$ of $\mathcal{O}_X$-modules, we have a natural isomorphism

$$\mathrm{Hom}(i(\mathcal{F}), \mathcal{P}) \xrightarrow{\simeq} \mathrm{Hom}(\mathcal{F}, \mathcal{P}^+)$$

which to each homomorphism $i(\mathcal{F}) \longrightarrow \mathcal{P}$ in the left-hand side group assigns the through map $i(\mathcal{F}) \longrightarrow \mathcal{P} \longrightarrow \mathcal{P}^+$ in the right-hand side group (the presheaf morphism $\mathcal{P} \longrightarrow \mathcal{P}^+$ is described in Sect. 2.1.5).

By means of the functors i and $+$ we can define the tensor operators over sheaves of modules. The general rule is as follows: perform the corresponding operation on presheaves and make the result into a sheaf with the help of $+$. In particular, given sheaves of $\mathcal{O}_X$-modules $\mathcal{F}_1, \mathcal{F}_2$ we define

$$\mathcal{F}_1 \otimes_{\mathcal{O}_X} \mathcal{F}_2 = (i(\mathcal{F}_1) \otimes_{\mathcal{O}_X} i(\mathcal{F}_2))^+,$$

and, for any sheaf morphism $f : \mathcal{F}_1 \longrightarrow \mathcal{F}_2$, we set

$$\mathcal{K}erf = (\mathcal{K}er\, i(f))^+ \quad \text{and} \quad \mathcal{C}okerf = (\mathcal{C}oker\, i(f))^+.$$

Actually, as is not difficult to show, $\mathrm{Ker}\, i(f)$ is automatically a sheaf, whereas for $\mathrm{Coker}\, i(f)$ this is not true, as Example 2.6.2 below shows. Somewhat later we will encounter examples in which $i(\mathcal{F}_1) \otimes_{\mathcal{O}_X} i(\mathcal{F}_2)$ is not a sheaf; these examples demonstrate the importance of the functor $+$.

Observe that i and $+$ are adjoint[4] functors.

2.6.1a Statement *The category of sheaves of modules is Abelian.*

The category of all sheaves of modules is usually too large. In what follows we will use two notions that single out a class of important sheaves needed in the sequel: quasi-coherent and coherent sheaves.

---

[4] Recall that an *adjunction between two categories* C and D is given by a pair of functors

$$F : D \longrightarrow C \text{ and } G : C \longrightarrow D$$

and a family of bijections

$$\mathrm{Hom}_C(FY, X) \cong \mathrm{Hom}_D(Y, GX)$$

which is natural for all variables $X \in C$ and $Y \in D$. The functor F is called a *left adjoint functor* to G, while G is called a *right adjoint functor* to F. For further details, see [GM].

2.6.2 Quasi-coherent Sheaves

A sheaf $\mathcal{F}$ of $\mathcal{O}_X$-modules is said to be *quasi-coherent* if it is locally isomorphic to the cokernel of a homomorphism of free sheaves.

More precisely, $\mathcal{F}$ is *quasi-coherent* if, for every $x \in X$, there exist a neighborhood $U \ni x$, two sets of indices I and J, and a homomorphism of sheaves of $\mathcal{O}_X|_U$-modules $f \colon \mathcal{O}_X^{(I)}|_U \longrightarrow \mathcal{O}_X^{(J)}|_U$, such that $\mathcal{F}|_U \simeq \mathcal{C}oker f$.

To elucidate the meaning of the quasi-coherence property, recall that the sheaves $\mathcal{O}_X^{(I)}$ correspond to "trivial bundles". The property of being isomorphic to the cokernel of a morphism of trivial bundles is a continuity-type condition: the jumps of the fibers should not be "too local", they should mirror a global picture over the open set.

2.6.2a Example Consider the case where X has the simplest non-trivial structure: $X = \operatorname{Spec} \mathbb{Z}_p$. Then $X = \{(0), (p)\}$, and the open sets are just $X, \{(0)\}$, and $\varnothing$, with the structure sheaf described by the following diagram, where $\mathbb{Z}_p \longrightarrow \mathbb{Q}_p$ is the natural embedding into the field of fractions:

$$
\begin{array}{ccc}
\Gamma(X, \mathcal{O}_X) & = & \mathbb{Z}_p \\
\downarrow & & \downarrow \\
\Gamma(\{(0)\}, \mathcal{O}_X) = & \mathbb{Q}_p &
\end{array}
$$

Any presheaf of modules over $(X, \mathcal{O}_X)$ is determined by a $\mathbb{Z}_p$-module F_1, a $\mathbb{Q}_p$-module F_2, and a $\mathbb{Z}_p$-homomorphism $F_1 \longrightarrow F_2$; moreover, every presheaf is a sheaf.

Since the only open neighborhood of the point (p) is X, it follows that the sheaf $\mathcal{F} = (F_1, F_2)$ is quasi-coherent if and only if there is an exact sequence of the form

$$
\mathcal{O}_X^{(I)} \longrightarrow \mathcal{O}_X^{(J)} \longrightarrow \mathcal{F} \longrightarrow 0.
$$

In terms of (F_1, F_2) this can be expressed as two exact sequences forming a commutative diagram:

$$
\begin{array}{ccccccc}
\mathbb{Z}_p^{(I)} & \longrightarrow & \mathbb{Z}_p^{(J)} & \longrightarrow & F_1 & \longrightarrow & 0 \\
\downarrow & & \downarrow & & \downarrow & & \\
\mathbb{Q}_p^{(I)} & \longrightarrow & \mathbb{Q}_p^{(J)} & \longrightarrow & F_2 & \longrightarrow & 0
\end{array}
$$

This immediately implies that $F_2 \simeq F_1 \otimes_{\mathbb{Z}_p} \mathbb{Q}_p$. It is not difficult to see that this condition is also sufficient for quasi-coherence of $\mathcal{F} = (F_1, F_2)$.

Thus, a quasi-coherent sheaf in this case is uniquely determined by the module of global sections $F_1 = \Gamma(X, \mathcal{O}_X)$, while F_2 and the homomorphism $F_1 \longrightarrow F_2$

are recovered from F_1. Without this quasi-coherence condition we have a greater freedom in defining both F_2 and the homomorphism $F_1 \otimes_{\mathbb{Z}_p} \mathbb{Q}_p \longrightarrow F_2$: now the sheaf may suffer a jump at a generic point as compared with the quasi-coherent case.

In Sect. 2.6.4 the result of this example will be extended to general affine schemes.

2.6.3 Coherent Sheaves

A restriction of "finite type" distinguishes *coherent* sheaves among the quasi-coherent ones.

A sheaf $\mathcal{F}$ of $\mathcal{O}_X$-modules is said to be a *sheaf of finite type* if it is locally isomorphic to the image of $\mathcal{O}_X^{(n)}$ for some $n \in \mathbb{N}$ under a homomorphism; in other words, for every $x \in X$, there exist an open neighborhood $U \ni x$ and a sheaf epimorphism $\mathcal{O}_X^{(n)}|_U \longrightarrow \mathcal{F}|_U \longrightarrow 0$. A sheaf of $\mathcal{O}_X$-modules is said to be *coherent* if it is of finite type and, for every open U and every sheaf morphism

$$\varphi \colon \mathcal{O}_X^{(n)}|_U \longrightarrow \mathcal{F}|_U \longrightarrow 0,$$

the sheaf $\mathcal{K}er\,\varphi$ over U is of finite type.

The general properties of coherent sheaves were first derived in Serre's thesis. We confine ourselves to listing them, cf. [KaS]:

(a) *A subsheaf of finite type of a coherent sheaf is coherent.*
(b) *If, in an exact sequence of sheaves*

$$0 \longrightarrow \mathcal{F} \longrightarrow \mathcal{G} \longrightarrow \mathcal{H} \longrightarrow 0,$$

   any two of the three sheaves are coherent, then the third one is also coherent. In particular, the direct sum of coherent sheaves and also the kernel, cokernel, and the image of any coherent sheaf under any morphism is coherent.
(c) *The tensor product of coherent sheaves is coherent.*
(d) *If the structure sheaf $\mathcal{O}_X$ is coherent, then a sheaf of $\mathcal{O}_X$-modules $\mathcal{F}$ is coherent if and only if it is locally isomorphic to the cokernel of a morphism of the form $\mathcal{O}_X^{(p)} \longrightarrow \mathcal{O}_X^{(q)}$.* □

2.6.4 Quasi-coherent Sheaves over Affine Schemes

Since the notion of quasi-coherence is local, it suffices to describe quasi-coherent sheaves over affine schemes. Let A be a ring and $(X, \mathcal{O}_X) = \operatorname{Spec} A$. Let M be an A-module. Our aim is to construct a sheaf $\widetilde{M}$ for which M is the module of global sections.

Set $M_x := M_{A \setminus \mathfrak{p}_x}$ and define $\Gamma(U, \widetilde{M})$ as the set of elements

$$\overline{m} := (\ldots, m_x, \ldots) \in \prod_{x \in U} M_x$$

such that for every $x \in U$, there exists a neighborhood of the form $D(f)$ and, for every $y \in D(f)$, the yth component m_y of $\overline{m}$ is the image of some $y \in M_f$ under the morphism $M_f \longrightarrow M_y$.

For example, $\widetilde{A} \simeq \mathcal{O}_X$.

2.6.5 Theorem $\Gamma(D(f), \widetilde{M}) \simeq M_f$, and the stalk of $\widetilde{M}$ over a point x is isomorphic to M_x.

Proof Define an isomorphism $\varphi: M_f \longrightarrow \Gamma(D(f), \widetilde{M})$ by setting

$$\varphi(m/f) = (\ldots, m/f, \ldots, m/f, \ldots) \in \prod_{x \in D(f)} M_x.$$

We skip the details that literally follow the proof of Theorem 2.2.4a. The natural map $M \longmapsto \widetilde{M}$ from the category of A-modules to the category of sheaves of $\mathcal{O}_X$-modules is actually a functor. Indeed, to every morphism $f: M \longrightarrow N$ there corresponds a morphism $\widetilde{f}: \widetilde{M} \longrightarrow \widetilde{N}$ which over each $D(f)$ is just the localization. The equality $\widetilde{fg} = \widetilde{f}\widetilde{g}$ is verified directly.   □

2.6.6 Proposition *For any exact sequence of A-modules $M \xrightarrow{f} N \xrightarrow{g} P$, the sequence of sheaves $\widetilde{M} \xrightarrow{f} \widetilde{N} \xrightarrow{g} \widetilde{P}$ is exact.*

Proof It suffices to verify the statement "stalk-wise" and apply Proposition 1.10.6a.
                                                                                    □

2.6.7 Proposition

(a) $M \cong \Gamma(\operatorname{Spec} A, \widetilde{M})$. (This means that not only $\widetilde{M}$ is recoverable from M, but also M is uniquely recoverable from $\widetilde{M}$.)
(b) $\operatorname{Hom}_A(M, N) \cong \operatorname{Hom}_{\mathcal{O}_X}(\widetilde{M}, \widetilde{N})$.

Proof (a) is a particular case of (b). To prove (b), observe that localization determines a natural map

$$\operatorname{Hom}_A(M, N) \longrightarrow \operatorname{Hom}_{\mathcal{O}_X}(\widetilde{M}, \widetilde{N}).$$

On the other hand, a morphism $\widetilde{M} \longrightarrow \widetilde{N}$ is a collection of morphisms $\widetilde{M}(U) \longrightarrow \widetilde{N}(U)$, among which there is a morphism

$$M \cong \Gamma(X, \widetilde{M}) \longrightarrow \Gamma(X, \widetilde{N}) \cong N.$$

This determines a map

$$\mathrm{Hom}_{\mathcal{O}_X}(\widetilde{M},\widetilde{N}) \longrightarrow \mathrm{Hom}_A(M,N).$$

The verification of the fact that the constructed maps are mutually inverse is trivial.

$\square$

2.6.7a Theorem *A sheaf $\mathcal{F}$ over $X = \mathrm{Spec}\,A$ is quasi-coherent if and only if $\mathcal{F} = \widetilde{M}$ for some A-module M.*

Proof

(a) Let us prove that if $\mathcal{F} = \widetilde{M}$, then $\mathcal{F}$ is quasi-coherent. Represent M as the cokernel of a free A-module morphism

$$A^{(I)} \longrightarrow A^{(J)} \longrightarrow M \longrightarrow 0.$$

This gives us the exact sequence of sheaves

$$\widetilde{A}^{(I)} \longrightarrow \widetilde{A}^{(J)} \longrightarrow \widetilde{M} \longrightarrow 0,$$

which implies the quasi-coherentness of $\widetilde{M}$, because $\widetilde{A} = \mathcal{O}_X$.

(b) Let $\mathcal{F}$ be quasi-coherent. Every point has a neighborhood of the form $D(f)$ over which $\mathcal{F}$ is isomorphic to the cokernel of a morphism of free sheaves. Let $\{D(f_i) \mid 1 \leq i \leq n\}$ be an open cover of $X = \mathrm{Spec}\,A$ by such neighborhoods, and let

$$\mathcal{F}|_{D(f_i)} = \mathrm{Coker}\left(\mathcal{O}_X^{(I)}|_{D(f_i)} \longrightarrow \mathcal{O}_X^{(J)}|_{D(f_i)}\right)$$

$$= \mathrm{Coker}\left(\widetilde{A}_{f_i}^{(I)} \longrightarrow \widetilde{A}_{f_i}^{(J)}\right) \simeq \mathrm{Coker}\left(A_{f_i}^{(I)} \longrightarrow A_{f_i}^{(J)}\right) = \widetilde{M}_i,$$

where M_i is an A_{f_i}-module. Notice that M_i can be also considered as an A-module (thanks to the localization homomorphism $A \longrightarrow A_{f_i}$). Further, let

$$M_{ij} = (M_i)_{f_j/1} = \Gamma(D(f_i f_j), \mathcal{F}) \quad \text{for } i, j = 1, \ldots, n.$$

Now, set

$$M = \mathrm{Ker}\left(\varphi\colon \prod_{1 \leq i \leq n} M_i \longrightarrow \prod_{1 \leq i, j \leq n} M_{ij}\right),$$

where φ is given by the formula

$$\varphi((\ldots, m_i, \ldots)) = (\ldots, m_{ij}, \ldots), \quad \text{where } m_{ij} = m_i/1 - m_j/1.$$

Let us prove that $\widetilde{M} \simeq \mathcal{F}$. It suffices to verify that

$$\Gamma(D(g), \mathcal{F}) = \Gamma(D(g), \widetilde{M}) \text{ for any } g \in A,$$

and this case easily reduces to $g = 1$ by replacing A with A_g and localizing all modules with respect to $\{g^n \mid n \in \mathbb{Z}_+\}$.

Therefore, it suffices to show that $\Gamma(\operatorname{Spec} A, \mathcal{F}) \simeq M$; but by the definition of the sheaf,

$$\Gamma(\operatorname{Spec} A, \mathcal{F}) = \operatorname{Ker}\left(\prod_{1 \leq i \leq n} \Gamma(D(f_i), \mathcal{F}) \longrightarrow \prod_{1 \leq i,j \leq n} \Gamma(D(f_i f_j), \mathcal{F})\right),$$

and by definition of $\widetilde{M}$ we have $\widetilde{M}|_{D(f_i)} = \widetilde{M}_i$ and $\widetilde{M}_{ij}|_{D(f_i f_j)} = \widetilde{M}_{ij}$. It only remains to apply Theorem 2.6.5.   □

2.6.8 Example

Let $(X, \mathcal{O}_X)$ be a scheme, $\mathcal{J}_X \subset \mathcal{O}_X$ a quasi-coherent sheaf of ideals. The quotient sheaf $\mathcal{O}_X/\mathcal{J}_X$ is obviously quasi-coherent. Define the *support of* $\mathcal{O}_X/\mathcal{J}_X$ by setting

$$\operatorname{supp} \mathcal{O}_X/\mathcal{J}_X := \{x \in X \mid \mathcal{O}_{X,x}/\mathcal{J}_{X,x} \neq \{0\}\}.$$

2.6.9 Lemma *If $\mathcal{J}_X$ is quasi-coherent, then* $\operatorname{supp} \mathcal{O}_X/\mathcal{J}_X$ *is closed in X and the ringed space* $(\operatorname{supp} \mathcal{O}_X/\mathcal{J}_X, \mathcal{O}_X/\mathcal{J}_X|_{\operatorname{supp} \mathcal{O}_X/\mathcal{J}_X})$ *is a scheme.*

Proof Consider an affine neighborhood $X \supset U \ni x$; i.e., let $U = \operatorname{Spec} A$. Denote $\mathcal{J} := \Gamma(U, \mathcal{J}_X) \subset A$. Obviously, $\operatorname{supp} \mathcal{O}_X/\mathcal{J}_X \cap U = V(F)$.   □

2.7 Invertible Sheaves and the Picard Group

How can one characterize the projective spectra of $\mathbb{Z}$-graded rings in intrinsic terms? The question is not very precise; a step towards its answer will be done in this section. We will show that the existence of a grading in the ring R enables one to define a particular quasi-coherent sheaf $\mathcal{O}_X(1)$ on $\operatorname{Proj} R = X$. We will introduce, certain invariants of R and show that they are actually characteristics of the pair $(X, \mathcal{O}_X(1))$.

2.7.1 Invertible Sheaves

A sheaf of modules L over a ringed space $(X, \mathcal{O}_X)$ is said to be *invertible*, if it is locally isomorphic, as a sheaf of $\mathcal{O}_X$-modules, to $\mathcal{O}_X$. The following statement is immediate:

2.7.1a Statement

(a) *If $\mathcal{L}_1$ and $\mathcal{L}_2$ are invertible sheaves, then the sheaf $\mathcal{L}_1 \otimes_{\mathcal{O}_X} \mathcal{L}_2$ is invertible.*
(b) *For any invertible sheaf $\mathcal{L}$ over $(X, \mathcal{O}_X)$, set*

$$\mathcal{L}^{-1} := \mathrm{Hom}_{\mathcal{O}_X}(\mathcal{L}, \mathcal{O}_X).$$

Then $\mathcal{L}^{-1} \otimes_{\mathcal{O}_X} \mathcal{L} \simeq \mathcal{O}_X$.

2.7.1a.i Corollary *The isomorphism classes of invertible sheaves over $(X, \mathcal{O}_X)$ constitute a commutative group with respect to tensoring over $\mathcal{O}_X$.*

This group is called the *Picard group* and denoted by $\mathrm{Pic}\,X$.

2.7.2 Cohomological Description of $\mathrm{Pic}\,X$

Let $\mathcal{L}$ be an invertible sheaf, $X = \bigcup_{i \in I} U_i$ an open cover of X sufficiently fine to satisfy $\mathcal{L}|_{U_i} \simeq \mathcal{O}_X|_{U_i}$ for all i. Fix an isomorphism $\varphi_i \colon \mathcal{L}_{U_i} \longrightarrow \mathcal{O}_X|_{U_i}$ and consider the restriction maps $r_{ij} \colon \mathcal{L}|_{U_i} \longrightarrow \mathcal{L}|_{U_i \cap U_j}$ and $r_{ji} \colon \mathcal{L}|_{U_j} \longrightarrow \mathcal{L}|_{U_i \cap U_j}$.

The isomorphisms φ_j are completely determined by the elements $\varphi_j^{-1}(1) = u_j \in \Gamma(U_j, \mathcal{L})$. Since $r_{ij}(u_i)$ and $r_{ji}(u_j)$ are generators of the module $\Gamma(U_i \cap U_j, \mathcal{L})$, it follows that the elements s_{ij}, determined from the equations $r_{ij}(u_i) = s_{ij} r_{ji}(u_j)$, are invertible, i.e.,

$$s_{ij} \in (\Gamma(U_i \cap U_j, \mathcal{O}_X))^{\times}.$$

Let $\Gamma(U, \mathcal{O}_X^{\times}) := \Gamma(U, \mathcal{O}_X)^{\times}$. Then, to a cover $(U_i)_{i \in I}$ of X, and an invertible sheaf $\mathcal{L}$ that is trivial on the elements of this cover, we assign the set

$$\{s_{ij} \in \Gamma(U_i \cap U_j, \mathcal{O}_X^{\times}) \mid i, j \in I\}.$$

Obviously, the elements s_{ij} satisfy the following conditions:

$$s_{ij} s_{ji} = 1, \quad \text{if } i \neq j,$$

$$s_{ij} s_{jk} s_{ki} = 1, \quad \text{if } i \neq j \neq k \neq i.$$

All such sets constitute a group with respect to multiplication, called the *group of 1-dimensional Čech cocycles of the cover* $(U_i)_{i \in I}$ *with coefficients in the sheaf* $\mathcal{O}_X^\times$ and denoted by $Z^1((U_i)_{i \in I}, \mathcal{O}_X^\times)$.

Two cocycles (s_{ij}), $(s'_{ij}) \in Z^1((U_i)_{i \in I}, \mathcal{O}_X^\times)$ are said to be *equivalent* if there exist $t_i \in \mathcal{O}_X^\times|_{U_i}$ such that $s'_{ij} = t_i s_{ij} t_j^{-1}$. The elements $t_i t_j^{-1}$ obviously constitute a subgroup $B^1((U_i)_{i \in I}, \mathcal{O}_X^\times) \subset Z^1$ of *coboundaries*.

The corresponding quotient group is called the *first Čech cohomology group of X with coefficients in* $\mathcal{O}^\times$ and denoted by $H^1((U_i)_{i \in I}, \mathcal{O}_X^\times)$.

The Čech cocycle constructed above for an invertible sheaf is multiplied by a coboundary if we change the isomorphisms φ_i. Indeed, let $\varphi'_i \colon \mathcal{L}|_{U_i} \xrightarrow{\sim} \mathcal{O}_X|_{U_i}$ be another set of isomorphisms. Since $\varphi_i {\varphi'_i}^{-1} \in \mathrm{Aut}(\mathcal{L}|_{U_i})$, we get $\varphi'_i = t_i \varphi_i$, whence $s'_{ij} = t_i s_{ij} t_j^{-1}$.

2.7.2a Proposition *The above map from the set of invertible sheaves* $\mathcal{L}$ *on X that are trivial on a given cover* $(U_i)_{i \in I}$ *into the first Čech cohomology set* $H^1((U_i)_{i \in I}, \mathcal{O}_X^\times)$ *determined by means of the same cover is one-to-one.*

2.7.2a.i Exercise Prove this proposition.

Thus, we obtained a group monomorphism

$$H^1((U_i)_{i \in I}, \mathcal{O}_X^\times) \longrightarrow \mathrm{Pic}\, X$$

whose image is the set of classes of sheaves trivial over all sets U_i.

Let $(U'_j)_{j \in J}$ be a finer cover. Then a natural monomorphism

$$H^1((U_i)_{i \in I}, \mathcal{O}_X^\times) \longrightarrow H^1((U'_j)_{j \in J}, \mathcal{O}_X^\times)$$

arises (we leave its precise construction to the reader). Since every invertible sheaf is trivial on elements of a sufficiently fine cover, we conclude that

$$\mathrm{Pic}\, X \simeq \varinjlim H^1((U_i)_{i \in I}, \mathcal{O}_X^\times) = H^1(X, \mathcal{O}_X^\times),$$

where the inductive limit is taken with respect to an ordered set of covers.

2.7.3   Example

On $X = \mathrm{Proj}\, R$, where R is generated by R_1 over R_0, consider a cover $X = \bigcup_{f \in R} U_f$, where $U_f = D_+(f)$, and a cocycle $s_{fg} \in Z^1(U_f, \mathcal{O}_X^\times)$ given by

$$s_{fg} = (f/g)^n \in \Gamma(U_f \cap U_g, \mathcal{O}_X^\times), \quad \text{where } n \in \mathbb{Z}.$$

The invertible sheaf determined by means of this cocycle is denoted by $\mathcal{O}_X(n)$; obviously, we have

$$\mathcal{O}_X(n) \simeq \begin{cases} \mathcal{O}_X(1)^{\otimes n}, & \text{if } n \geq 0, \\ \mathcal{O}_X(-1)^{\otimes n}, & \text{if } n \leq 0, \end{cases} \tag{2.5}$$

where $\mathcal{O}_X(-1) = \mathcal{O}_X(1)^{-1}$.

These sheaves are constructed from R; the other way around, R can be recovered to some extent from X and $\mathcal{O}_X(1)$. Here we will only prove a part of the result; the second part will be proved with the help of the cohomology technique in what follows.

First, notice that for every invertible sheaf $\mathcal{L}$ over a ringed space $(X, \mathcal{O}_X)$, there is a natural structure of a $\mathbb{Z}$-graded ring on $\bigoplus_{n \in \mathbb{Z}} \Gamma(X, \mathcal{L}^n)$ with the product of homogeneous elements determined by means of the map

$$\Gamma(X, \mathcal{L}^n) \times \Gamma(X, \mathcal{L}^m) \longrightarrow \Gamma(X, \mathcal{L}^n \otimes_{\mathcal{O}_X} \mathcal{L}^m) \cong \Gamma(X, \mathcal{L}^{n+m}).$$

2.7.4 Theorem *Let* $X = \text{Proj } R$, *where* R_0 *is a Noetherian ring and* R_1 *is a Noetherian* R_0-*module that generates* R *over* R_0, *and* $\mathcal{L} = \mathcal{O}_X(1)$. *Then there exist a homogeneous homomorphism of graded rings*

$$\alpha : R \longrightarrow \bigoplus_{n \in \mathbb{Z}} \Gamma(X, \mathcal{L}^n),$$

and an $n_0 \in \mathbb{Z}$ *such that the maps* $\alpha_n : R_n \longrightarrow \Gamma(X, \mathcal{L}^n)$ *are group isomorphisms for all* $n \geq n_0$.

Proof Let us construct α and show that its kernel is only supported in small degrees. The fact that the α_n are isomorphism will be proved in what follows.

Let $h \in R_n$. For any $f \in R_1$, set

$$\alpha(h)|_{D_+(f)} = \frac{h}{f^n} \in \Gamma(D_+(f), \mathcal{O}_X).$$

Obviously, the sections h/f^n are glued together with the help of the cocycle $(f/g)^n$ into a section of $\mathcal{L}^n$ over the whole X; denote this section by $\alpha(h)$. Clearly, α is a homomorphism of graded rings.

Let $h \in R_n \cap \text{Ker}\,\alpha$. This means that $h/f^n = 0$ for all $f \in R_1$. Since R_1 is Noetherian, there exists an integer m_0 such that $R_m h = 0$ for $m \geq m_0$. Consider an arbitrary h whose annihilator contains $\bigoplus_{m \geq m_0(h)} R_m$. Obviously, all such elements constitute an ideal $J \subset R$.

Since R is Noetherian, J has finitely many generators, and therefore we can choose one m_0 for all generators h; hence $J_m = 0$ for $m \geq m_0$. $\qquad \square$

2.7.5   Picard Group: Examples

2.7.5a Proposition *Let A be a unique factorization ring.*
   Then $\mathrm{Pic}(\mathrm{Spec}\, A) = \{0\}$.

Proof In the set of all open covers, the finite covers of the form $\bigcup_{i \in J} D(f_i)$ constitute a *cofinal subsystem*,[5] and therefore it suffices to verify that

$$H^1(D(f_i), \mathcal{O}_X^\times) = \{0\}.$$

Let $s_{ij} \in Z^1(D(f_i), \mathcal{O}_X^\times)$. Let us represent all s_{ij} for $i \neq j$ in the form t_i'/t_j', where the t_i' are elements of the field of fractions K of A. It is easy to see that this is indeed possible: because $s_{ij}s_{jk}s_{ki} = 1$ for any k, it follows that $s_{ij} = s_{ik}/s_{jk}$, since $s_{ik}s_{ki} = 1$.

Now, let p be a prime element of A and $v_p(a)$ the exponent with which p enters the decomposition of $a \in K$. Up to multiplication by invertible elements, the set P of primes $p \in A$ such that $v_p(t_i') \neq 0$ for some i is finite.

Fix $p \in P$, and divide all f_i into two groups: with p dividing f_i for $i \in J_1$ and with p not dividing f_i for $i \in J_2$. Since the f_i are relatively prime, $J_2 \neq \emptyset$.

Since s_{ij} is invertible in $A_{f_if_j}$, we see that $v_p(s_{ij}) = 0$ if p does not divide f_{ij}. Therefore, $v_p(t_i')$ takes the same value, denoted by a_p, for all $i \in J_2$. Set

$$t_i = \left(\prod_{p \in P} p^{-a_p} \right) t_i'.$$

Obviously, $s_{ij} = t_i/t_j$; on the other hand, $t_i \in \Gamma(D(f_i), \mathcal{O}_X) = A_{f_i}^\times$. Indeed, if p does not divide f_i, then $v_p(t_i) = 0$ and t_i factorizes into the product of the prime divisors of f_i, which implies that t_i is invertible in A_{f_i}.   □

2.7.5b Remark In terms of cohomology with coefficients in sheaves, we can interpret this proof as follows. The exact sequence of sheaves of Abelian groups on X

$$1 \longrightarrow \mathcal{O}_X^\times \longrightarrow \widetilde{K}^\times \stackrel{p}{\longrightarrow} \widetilde{K}^\times/\mathcal{O}_X^\times \longrightarrow 1,$$

where $\widetilde{K}^\times$ is the constant sheaf (i.e., $\Gamma(U, \widetilde{K}^\times) = K^\times$ for any U), induces the exact sequence of cohomology groups

$$\Gamma(X, \widetilde{K}^\times) \stackrel{p_0^\times}{\longrightarrow} \Gamma(X, \widetilde{K}^\times/\mathcal{O}_X^\times) \longrightarrow H^1(X, \mathcal{O}_X) \stackrel{p_1^\times}{\longrightarrow} H^1(X, \widetilde{K}^\times).$$

---

[5]Recall that a subset B of a partially ordered set A is said to be *cofinal* if, for every $a \in A$, there exists $b \in B$ such that $a \leq b$. A sequence or net of elements of A is said to be *cofinal* if its image is cofinal in A.

   In our situation, we require that every cover has a *finite refinement* such that every open of the refinement lies in an open of the initial cover.

The first step of the above proof establishes that $p_1^\times = 0$ (actually, the same argument shows that even $H^1(X, \widetilde{K}^\times) = 0$). The second step shows that $p_0^\times$ is an epimorphism; and it is only here that we have used the fact that A is a unique factorization ring, which, in particular, implies that $\widetilde{K}^\times / \widetilde{A}^\times \simeq \bigoplus_{p \in \operatorname{Spec} A} \widetilde{\mathbb{Z}}$.

Here is an important application of the above Proposition 2.7.5a:

2.7.5c Theorem *Let A be a unique factorization ring. Then* $\operatorname{Pic} \mathbb{P}_A^r$ *for $r \geq 1$ is an infinite cyclic group with the class of $\mathcal{O}(1) := \mathcal{O}_X(1)$, where $X = \mathbb{P}_A^r$, as its generator.*

Proof Recall that $\mathbb{P}_A^r = \operatorname{Proj} A[T_0, \ldots, T_r]$. By Proposition 2.7.5a, any invertible sheaf $\mathcal{L}$ over $\mathbb{P}_A^r$ is trivial on $D_+(T_i) = \operatorname{Spec} A\left[\dfrac{T_0}{T_i}, \ldots, \dfrac{T_r}{T_i}\right]$, because by a theorem of Gauss (see, e.g., [Pr]) *the polynomial ring over A inherits the unique factorization property of A.*

Now, let $(s_{ij} \mid 0 \leq i, j \leq r)$ be a cocycle defining $\mathcal{L}$ for the cover $(D_+(T_i) \mid 0 \leq i \leq r)$. Since s_{ij} is homogeneous of degree 0 and only factorizes in the product of the divisors of $T_i T_j$ (use the unique factorization property of $A[T_0, \ldots, T_r]$), we have

$$s_{ij} = \varepsilon_{ij} \left(\frac{T_i}{T_j}\right)^{n_{ij}}, \quad \text{where } \varepsilon_{ij} \in A^\times. \tag{2.6}$$

Since $s_{ij} s_{ji} = 1$ and $s_{ij} s_{jk} s_{ki} = 1$, it follows that $n_{ij} = n$ (does not depend on i, j), and therefore ε_{ij} is a cocycle.

In actual fact, ε_{ij} is automatically a coboundary, because $\varepsilon_{ij} = \varepsilon_{ik}/\varepsilon_{jk}$ for any k and $\varepsilon_{ik} \in \Gamma(\mathbb{P}_A^r, \mathcal{O}_X^\times)$. Therefore, (s_{ij}) is cohomologous to the cocycle $(T_i/T_j)^n$ defining $\mathcal{O}(n)$.

We will now get the statement of the theorem if we prove that all sheaves $\mathcal{O}(n)$ are nonisomorphic. This is true for any A, as shown by the following result:

2.7.5d Lemma *Let $\mathbb{P}_A^r = \operatorname{Proj}(A)[T_0, \ldots, T_n]$ for any A. Then*

$$\Gamma(\mathbb{P}_A^r, \mathcal{O}(n)) = \begin{cases} 0, & \text{if } n < 0, \\ \displaystyle\bigoplus_{a_0 + \cdots + a_r = n} A T_0^{a_0} \cdots T_r^{a_r}, & \text{if } n \geq 0. \end{cases}$$

Proof Let $R = A[T_0, \ldots, T_r]$, and $\deg T_i = 1$ for all i. Let us show that the homomorphism $\alpha_n: R_n \longrightarrow \Gamma(\mathbb{P}_A^r, \mathcal{O}(n))$ is an isomorphism for $n \geq 0$.

Recall that $\alpha_n(f)|_{D_+(T_i)} = f/T_i^n$ for any $f \in R_n$. Now, the fact that the T_i are not zero divisors immediately implies that α_n is a monomorphism.

Let us prove that α_n is an epimorphism. A section of the sheaf $\mathcal{O}_X^\times$ over $\mathbb{P}_A^r$ is represented by a collection

$$\left\{ f_i \in A\left[\frac{T_0}{T_i}, \ldots, \frac{T_r}{T_i}\right] \mid 0 \leq i \leq n, \text{ and } f_i \left(\frac{T_i}{T_j}\right)^n = f_j \right\}.$$

Since the T_i are not zero divisors, the compatibility conditions imply that $f_i T_i^n$ do not depend on i. Obviously, $f_i T_i^n$ is a polynomial because its denominator can only be a power of T_i.

Now, let $n < 0$; then, with the same notation, we get

$$f_i / T_i^{-n} = f_j / T_j^{-n},$$

and by similar divisibility considerations this is only possible for $f_i = 0$. $\square$

The theorem is thus also proved. $\square$

2.7.5e Corollary $\mathcal{O}(n) \not\cong \mathcal{O}(m)$ *if* $n \neq m$.

Proof This immediately follows from Lemma 2.7.5d because the ranks of the A-modules of sections of a certain power of these sheaves are distinct. $\square$

2.7.6   Hilbert Polynomial of the Projective Space

As an application of Theorems 2.7.5c and 2.7.4 we can now compute the Hilbert polynomials of projective spaces over a given field and determine which of the numerical characteristics of $\mathbb{P}_R^r$ introduced in Sect. 2.5 do not depend on the representation of $\mathbb{P}_R^r$ in the form $\operatorname{Proj} R$.

Indeed, let $\mathbb{P}_k^r = \operatorname{Proj} R$; temporarily, denote by $\mathcal{O}_R(1)$ the invertible sheaf on $\mathbb{P}_k^r$ constructed with the help of the ring R, and let $\mathcal{O}(1)$ be the invertible sheaf constructed by means of the standard representation $\mathbb{P}_k^r = \operatorname{Proj} k[T_0, \dots, T_r]$.

By Theorem 2.7.5c,

$$\mathcal{O}_R(1) \simeq \mathcal{O}_R(d) \ \text{ for some } d \in \mathbb{Z};$$

for $r \geq 1$, we have $d > 0$ because the rank of the space of sections of the sheaf $\mathcal{O}_R(n)$ grows as $n \longrightarrow \infty$.

On the other hand, by a (not yet proved!) part of Theorem 2.7.4 for n sufficiently large, the map

$$\alpha_n : R_n \longrightarrow \Gamma(\mathbb{P}_k^r, \mathcal{O}_k(n)) = \Gamma(\mathbb{P}_k^r, \mathcal{O}(nd))$$

is an isomorphism. Hence,

$$h_R(n) = \binom{nd + r}{r}.$$

In particular, the *degree* and the *constant term* of the Hilbert polynomial do not depend on R, as claimed.

2.7.7 Exercises

(1) Prove that a curve in $\mathbb{P}^r_k$ can be isomorphic to the projective line $\mathbb{P}^1_k$ only if its degree is equal to 1 or 2.

Hint By definition, any curve in $\mathbb{P}^r_k$ is of the form $\operatorname{Proj} k[T_0, T_1, T_2]/(f)$, where f is a form. Its Hilbert polynomial is computed in Sect. 2.5.

Try to prove that the curve determined by a quadratic form f is isomorphic to $\mathbb{P}^1_k$ if and only if the following two conditions are fulfilled: (1) $\operatorname{rank} f = 3$; (2) the equation $f = 0$ has a non-zero solution in k.

(2) Let $r \geq 1$; prove that any automorphism $f: \mathbb{P}^r_k \longrightarrow \mathbb{P}^r_k$ over k is linear (what does this mean?).

Hint Consider how f acts on invertible sheaves and on the Picard group.

2.8 The Čech Cohomology

2.8.1 The Čech Complex

Let X be a topological space and $\mathcal{F}$ a sheaf of Abelian groups on X. Let $U = \bigcup_{i=1}^{r} U_i$ be a finite open cover of X. A *Čech complex* is a complex whose homogeneous components $C^p(U, \mathcal{F})$ (called *groups of Čech p-cochains*) and differential are defined as follows:

Let $[1, r]^{p+1}$ be the $(p+1)$-fold direct product of the set of integers $1, \ldots, r$. The elements of $C^p(U, \mathcal{F})$ are the functions

$$s(i_0, \ldots, i_p) \in \Gamma(U_{i_0 \ldots i_p}, \mathcal{F}), \quad \text{where } U_{i_0 \ldots i_p} := U_{i_0} \cap \ldots \cap U_{i_p},$$

"antisymmetric" in the sense that

$$s(\sigma(i_0), \ldots, \sigma(i_p)) = \operatorname{sgn} \sigma \cdot s(i_0, \ldots, i_p) \quad \text{for any permutation } \sigma \in S_p$$

and hence

$$s(i_0, \ldots, i_p) = 0 \quad \text{if among the indices } i_0, \ldots, i_p \text{ at least two coincide.}$$

In particular, $C^p(U, \mathcal{F}) = 0$ for $p \geq r$; moreover, $C^0(U, \mathcal{F}) = \prod_{i=1}^{r} \Gamma(U_i, \mathcal{F})$.

The differential in the Čech complex acts as (**hereafter in similar sums, the hatted symbol should be ignored** together with commas around it, if any):

$$(ds)(i_0, \ldots, i_{p+1}) = \sum_{k=0}^{p+1} (-1)^k rs(i_0, \ldots, \widehat{i_k}, \ldots, i_{p+1}),$$

where $r: \Gamma(U_{i_0 \dots \widehat{i_k} \dots i_{p+1}}, \mathcal{F}) \longrightarrow \Gamma(U_{i_0 \dots i_{p+1}}, \mathcal{F})$ is the restriction homomorphism (which, obviously, depends on $i_0, \dots, i_{p+1}$ and k).

The cohomology groups of this complex are called the *Čech cohomology groups of the cover U with coefficients in the sheaf* $\mathcal{F}$ and are denoted $\check{H}^p(U, \mathcal{F})$. Cohomology with coefficients in a sheaf defined à la Čech are convenient for calculations. However, they characterize, in a sense, the cover U of the space X and sections of the sheaf $\mathcal{F}$ over its charts, rather than the space X on which the sheaf $\mathcal{F}$ is given.

Remark Grothendieck suggested an axiomatic definition of cohomology of the space X with coefficients in a sheaf $\mathcal{F}$. According to his definition, *the pth cohomologies $H^p(X, \mathcal{F})$ of X with coefficients in a sheaf $\mathcal{F}$ are the right derived functors* (see footnote 4 on p. 141) of the functor that to every sheaf of Abelian groups $\mathcal{F}$ on X assigns its group of sections $\Gamma(X, \mathcal{F})$.

Let us formulate without proof a theorem offering a sufficient condition for coincidence of Grothendieck's $H^p(X, \mathcal{F})$ with Čech's $\check{H}^p(U, \mathcal{F})$.

2.8.2 Theorem (Cartan) *Let $\mathcal{V}$ be a family of quasi-compact open subsets of a topological space X forming a basis of the topology of X and such that $\check{H}^p(U, \mathcal{F}) = 0$ for all $p \geq 1$ and $\bigcup_{i=1}^r U_i \in \mathcal{V}$ for all finite covers $U = (U_i)_{i=1}^r$ such that $U_i \in \mathcal{V}$. Then $\check{H}^\cdot(X, \mathcal{F}) \cong H^\cdot(X, \mathcal{F})$ for any finite cover U of X by elements of $\mathcal{V}$.*

2.8.2i Remark The theorem is not formulated in full generality, but it suffices for our nearest purposes. For its proof, see [God, Theorem 5.9.2 in Ch. 2].

We will apply this theorem to schemes X taking the families of affine open sets as $\mathcal{V}$ and taking any quasi-coherent sheaf as $\mathcal{F}$. Let us establish that in this situation the conditions of Cartan's theorem are fulfilled. For this, it suffices to prove the following result.

2.8.3 Proposition *Let $X = \operatorname{Spec} A$ and $\mathcal{F} = \widetilde{M}$, where M is an A-module. Let $U_i = D(f_i)$, where $i = 1, \dots, r$; let $X = \bigcup U_i$, and $U = (U_i)_{i=1}^r$. Then*

$$\check{H}^p(U, \mathcal{F}) = 0 \text{ for } p \geq 1.$$

2.8.3a Corollary *For any affine scheme X and any quasi-coherent sheaf $\mathcal{F}$ on X, we have $H^p(X, \mathcal{F}) = 0$ for any $p \geq 1$.*

Proof Apply Cartan's theorem to X and the family of opens $D(f)$.   □

2.8.3b Corollary *For any separable[6] scheme X, any finite cover U by affine schemes, and any quasi-coherent sheaf $\mathcal{F}$ on X, we have*

$$\check{H}^p(U, \mathcal{F}) = H^p(X, \mathcal{F}).$$

---

[6]The separability condition implies that the intersection of two affine open subsets is affine; the schemes Proj are always separable.

Proof Apply Cartan's theorem to X and the family of affine open subsets of X using Corollary 2.8.3a.   □

We note, without proof, that Serre proved the converse of Corollary 2.8.3a:

if for a scheme X and any quasi-coherent sheaf of ideals $\mathfrak{J}$ on X, we have $H^1(X, \mathfrak{J}) = 0$, then X is an affine scheme.

2.8.4 Properties of Čech Complexes

We retain the notation of Proposition 2.8.3. Let

$$U_{i_0 \ldots i_p} = U_{i_0} \cap \ldots \cap U_{i_p} = D(f_{i_0} \cdots f_{i_p}).$$

We have

$$\Gamma(U_{i_0 \ldots i_p}, \widetilde{M}) \simeq M_{f_{i_0} \cdots f_{i_p}}.$$

Each cochain of $C^p(U, \widetilde{M})$ with $p < r$ can be represented by a collection of $\binom{r}{p+1}$ elements of different localizations of the module M,

$$s(i_0, \ldots, i_p) \in M_{f_{i_0} \cdots f_{i_p}}.$$

We wish to prove that the Čech complex is *acyclic* in dimensions $p \geq 1$, i.e., the cohomologies of this complex are 0 in these dimensions. The standard method of proving acyclicity is to construct a *homotopy operator* or, more precisely, a sequence of *chain homotopy operators*, thereby establishing the equivalence of the given complex and an acyclic one.[7] For the Čech complex we cannot construct such an

---

[7]Let (A, d) and (A', d') be chain complexes and $f : A \longrightarrow A'$, $g : A \to A'$ be chain maps. A *chain homotopy* D between f and g is a sequence of homomorphisms $\{D_n : A_n \longrightarrow A'_{n+1}\}$ such that

$$d'_{n+1} \circ D_n + D_{n-1} \circ d_n = f_n - g_n \text{ for each } n.$$

Thus, we have the diagram

$$
\begin{array}{ccccc}
A_{n+1} & \xrightarrow{d_{n+1}} & A_n & \xrightarrow{d_n} & A_{n-1} \\
\downarrow{\scriptstyle f_{n+1}-g_{n+1}} & \stackrel{D_n}{\nearrow} & \downarrow{\scriptstyle f_n-g_n}{\scriptstyle D_{n-1}} & \stackrel{}{\nearrow} & \downarrow{\scriptstyle f_{n-1}-g_{n-1}} \\
A'_{n+1} & \xrightarrow{d'_{n+1}} & A'_n & \xrightarrow{d'_n} & A'_{n-1}
\end{array}
$$

If there exists a chain homotopy between f and g, then f and g are said to be *chain homotopic*. Any complex that is chain homotopic to one with zero cohomology is said to be *acyclic*.

operator, but we will circumvent this difficulty as follows. We construct a sequence of complexes $(C_n^p(M))$, where n is the number of the complex, and homomorphisms between them, such that:

(a) *The Čech complex is the inductive limit of the complexes $C_n^p(M)$.*

(b) *The complexes $C_n^p(M)$ are acyclic.*

$$(2.7)$$

Since the passage to the inductive limit commutes with taking the cohomology, we see that

$$\check{H}^p(U, \widetilde{M}) = H(C^p(U, \widetilde{M})) = 0.$$

2.8.4a   Step (a) of the Program (2.7)

First, let us prove that for any ring A, any A-module M, and any $g \in A$, the module M_g can be naturally represented as an inductive limit.

First of all, by setting

$$M_g^{(n)} = \left\{ \frac{m}{g^n} \mid m \in M \right\},$$

we see that $M_g = \bigcup_{n=0}^{\infty} M_g^{(n)}$. Each space $M_g^{(n)}$ is an A-module and we can replace the union with the inductive limit by considering the system

$$\cdots \longrightarrow M_g^{(n)} \longrightarrow M_g^{(n+1)} \longrightarrow \cdots , \qquad (2.8)$$

in which the homomorphisms are given by the rule $\dfrac{m}{g^n} \longmapsto \dfrac{mg}{g^{n+1}}$. If g is not a zero divisor in M, then the A-module $M_g^{(n)}$ is isomorphic to M for all n via the map $m \longmapsto \dfrac{m}{g^n}$. The inductive system (2.8) turns now into

$$\cdots \xrightarrow{g} M^{(n)} = M \xrightarrow{g} M^{(n+1)} = M \xrightarrow{g} \cdots , \qquad (2.9)$$

where each slot is occupied by a copy of the A-module M, and each homomorphism is multiplication by g.

2.8.4a.i. Lemma *The inductive limit of the system (2.9) is always isomorphic to M_g (even if g is a zero divisor in M).*

Proof Consider the homomorphisms

$$M^{(n)} \longrightarrow M_g, \quad m \longmapsto \frac{m}{g^n}.$$

They are compatible with the homomorphisms of the system (2.9) and therefore define a homomorphism of its limit,

$$\varinjlim M^{(n)} \longrightarrow M_g.$$

Its cokernel is, clearly, zero. Indeed, any element of the kernel is represented by a chain of elements $gm_n, g^2 m_n, \ldots$, where $m_n \in M^{(n)} = M$, such that $m_n/g^n = 0$ in M_g; this means that $g^{n+k} m_n = 0$ for k sufficiently large, and therefore the whole chain represents the zero class. □

Now, in the whole complex $(C^p(U, \widetilde{M}))$, replace the localizations of M with their "approximations" $M^{(n)}$ and appropriately define the cochain operators.

To make the correct expression of the cochains graphic, we first assume that the elements $f_{i_0}, \ldots, f_{i_p}$ are not zero divisors in M. Let $C_n^p(M)$ denote the subgroup of $C^p(U, \widetilde{M})$ consisting of the cochains

$$s(i_0, \ldots, i_p) \in M^{(n)}_{f_{i_0} \cdots f_{i_p}} = \left\{ \frac{m}{(f_{i_0} \cdots f_{i_p})^n} \mid m \in M \right\}.$$

Let

$$s(i_0, \ldots, i_p) = \frac{m_{i_0 \ldots i_p}}{(f_{i_0} \cdots f_{i_p})^n}.$$

Then

$$ds(i_0, \ldots, i_{p+1}) = \frac{m_{i_0 \ldots i_{p+1}}}{(f_{i_0} \cdots f_{i_{p+1}})^n} = \sum_{k=0}^{p+1} (-1)^k \frac{m_{i_0 \ldots \widehat{i_k} \ldots i_{p+1}}}{(f_{i_0} \cdots \widehat{f}_{i_k} \cdots f_{i_{p+1}})^n}, \qquad (2.10)$$

whence

$$m_{i_0 \ldots i_{p+1}} = \sum_{k=0}^{p+1} (-1)^k f_{i_k}^n m_{i_0 \ldots \widehat{i_k} \ldots i_{p+1}}. \qquad (2.11)$$

The embedding homomorphism $C_n^p(M) \longrightarrow C_{n+1}^p(M)$ is described by the formula

$$m_{i_0 \ldots i_p} \longmapsto f_{i_0} \cdots f_{i_p} m_{i_0 \ldots i_p}. \qquad (2.12)$$

We use formulas (2.11) and (2.12) to define both the differential in the complex $C_n^p(M)$ when the condition on zero divisors is not satisfied, and the homomorphism of complexes $C_n^p(M) \longrightarrow C_{n+1}^p(M)$.

In the general case, denote by $C_n^p(M)$ the group of antisymmetric functions on $[1, r]^{p+1}$ with values in M and define the coboundary operator $C_n^p(M) \longrightarrow C_n^{p+1}(M)$ by setting:

$$(dm)(i_0, \ldots, i_{p+1}) = \sum_{k=0}^{p+1} (-1)^k f_{i_k}^n m(i_0, \ldots, \widehat{i_k}, \ldots, i_{p+1}).$$

Define the group homomorphism $\varphi_n = C_n^p(M) \longrightarrow C_{n+1}^p(M)$ by the formula

$$(\varphi_n m)(i_0, \ldots, i_p) = f_{i_0} \cdots f_{i_p} m(i_0, \ldots, i_p).$$

2.8.4a.ii. Lemma

(1) *The collection of sets* $(C_n^p(M))$ *is a complex for every fixed n.*
(2) *The collection of homomorphisms* φ_n *is a homomorphism of complexes.*
(3) $\varinjlim C_n^p(M) = C^p(U, \widetilde{M})$; *the inductive limit of differentials is the differential in the inductive limit.*

Proof The first two statements are verified by trivial calculations; the third one follows from definitions and Lemma 2.8.4a.ii.   □

This concludes step (a) in computing cohomology of the Čech complex, i.e., approximating it by complexes $C_n^p(M)$ which are easier to deal with.

2.8.4b   Step (b) of the Program (2.7)

Let us now prove that the complex $C_n^p(M)$ is acyclic. The construction of this complex involves the ring A, the A-module M, the elements $f_1^n, \ldots, f_r^n \in A$ that determine the cover $\{D(f_i^n)\}_{i=1}^r = \{D(f_i)\}_{i=1}^r$, and the differential given by formula (2.10).

Since the complex $C_n^p(M)$ is important in various problems of algebraic geometry, we will study it in more detail than is actually necessary for our purposes.

2.8.5   The Koszul Complex

Let A be a ring, $f = (f_1, \ldots, f_r)$ a collection of its elements; set $f^n := (f_1^n, \ldots, f_r^n)$.

Consider a free A-module $Ae_1 \oplus \cdots \oplus Ae_r = A^r$ of rank r and its exterior powers $K_p = \bigwedge_A^p(A^r)$; by definition, $K_0 = A$. Clearly, K_p is a free A-module of rank $\binom{r}{p}$; the elements $e_{i_1} \wedge \cdots \wedge e_{i_p}$, where $i_1 < \cdots < i_p$, form a basis in it. Define the differential $d: K_{p+1} \longrightarrow K_p$ by setting

$$d(e_{i_1} \wedge \cdots \wedge e_{i_{p+1}}) = \sum_{k=1}^{p+1} (-1)^{k+1} f_{i_k} e_{i_1} \wedge \cdots \wedge \widehat{e}_{i_k} \wedge \cdots \wedge e_{i_{p+1}}.$$

(**NB**: We raised -1 to the power $k + 1$ in order for the first term to enter with its initial sign.) It is a trivial job to verify that $d^2 = 0$; let $K_p(f, M)$ or briefly $K_p(f)$ be the resulting complex; it is called the *Koszul complex*.

Observe that the Koszul complex is a *chain* complex, whereas the Čech complex is a *cochain* complex. The complexes $K_p(f)$ and $C_n^p(M)$ are related as follows.

2.8.5a Lemma *For $p \geq 0$, we have*

$$C_n^p(M) \simeq K^{p+1}(f^n, M) := \mathrm{Hom}(K_{p+1}(f^n), M),$$

and the isomorphism can be selected to be compatible with the differentials.

Proof To any cochain $m = (m(i_0, \ldots, i_p)) \in C_n^p(M)$ we assign the homomorphism

$$K_{p+1}(f^n) \longrightarrow M,$$

$$g_m \colon e_{i_0} \wedge \cdots \wedge e_{i_p} \longmapsto m(i_0, \ldots, i_p).$$

The differential of g_m considered as an element of $\mathrm{Hom}(K_{p+2}(f^n), M)$, is given by the formula

$$(dg_m)(e_{i_0} \wedge \cdots \wedge e_{i_{p+1}}) = g_m(d(e_{i_0} \wedge \cdots \wedge e_{i_{p+1}}) =$$

$$= g_m \left(\sum_{k=0}^{p+1} (-1)^k f_{i_k}^n m(i_0, \ldots, \widehat{i}_k, \ldots, i_{p+1}) \right) = g_{dm}(e_{i_0} \wedge \cdots \wedge e_{i_{p+1}}),$$

which proves the lemma.   □

Thus, we have exhibited the dependence of $C_n^p(M)$ on M. The complex $K_p(f)$ can also be further "dismantled"; this is convenient in proofs based on induction on r.

2.8.5b Lemma $K_0(f) \simeq K_0(f_1) \otimes \cdots \otimes K_0(f_r)$.

Proof First of all, recall that the *tensor product of two chain complexes K_0 and L*, is the complex such that

$$(K \otimes L)_p = \bigoplus_{i+j} K_i \otimes L_j,$$

$$d(k \otimes l) = dk \otimes l + (-1)^r k \otimes dl, \text{ where } k \in K_i;$$

and, generally, for $K^{(1)} \otimes \cdots \otimes K^{(r)}$, we have

$$d(k_1 \otimes \cdots \otimes k_r) = \sum_{j=1}^{r} (-1)^{d_1 + \cdots + d_{j-1}} k_1 \otimes \cdots \otimes \widehat{k_j} \otimes \cdots \otimes k_r,$$

$$\text{where } k_j \in K_{d_j}^{(j)} \text{ for } j = 1, \ldots, r.$$

Let now $K_0^{(i)} = A$ and $K_1^i = Ae_i$ for all i. Construct a complex by setting

$$0 \longleftarrow A \longleftarrow Ae_i, \quad d(e_i) = f_i.$$

Then $(K^{(1)} \otimes \cdots \otimes K^{(r)})_p$ is a free A-module with a basis $e_{i_1} \otimes \cdots \otimes e_{i_p}$, where $1 \le i_1 < \cdots < i_r \le p$, and the differential

$$d(e_{i_1} \otimes \cdots \otimes e_{i_p}) = \sum_{k=1}^{p} (-1)^{k+1} f_{i_k} e_{i_1} \otimes \cdots \otimes \widehat{e_{i_k}} \otimes \cdots \otimes e_{i_p}.$$

This shows that it is isomorphic to the Koszul complex $K_0(f)$. We can now prove that the complex $C_n^p(M)$ is acyclic in dimensions ≥ 1. For this, it suffices to construct a homotopy for complexes $K_p(f)$. $\qquad\qquad\qquad\square$

2.8.5c Proposition *Let $g_1, \ldots, g_r$ be an arbitrary collection of r elements of A. Let $h: K_p(f) \longrightarrow K_{p+1}(f)$ be the homomorphism of exterior multiplication on the left by $\sum_{i=1}^{r} g_i e_i$. Then*

$$hd + dh = \sum_{i=1}^{r} f_i g_i$$

(i.e., the homomorphism of multiplication by the sum on the right).

Proof Fix a set of indices (without repetitions) $i_1, \ldots, i_{p+1} \in [1, r]$ and let $j_1, \ldots, j_{r-p-1}$ be the complementary set of indices. We have:

$$dh(e_{i_1} \wedge \cdots \wedge e_{i_{p+1}}) =$$

$$= d\left(\left(\sum_{k=1}^{r} g_k e_k \right) \wedge e_{i_1} \wedge \cdots \wedge e_{i_{p+1}} \right) = d\left(\sum_{k=1}^{r-p-1} g_{j_k} e_{j_k} \wedge e_{i_1} \wedge \cdots \wedge e_{i_{p+1}} \right) =$$

$$\sum_{k=1}^{r-p-1} (g_{j_k} f_{j_k} e_{i_1} \wedge \cdots \wedge e_{i_{p+1}}) + g_{j_k} \sum_{l=1}^{p+1} (-1)^l f_{i_l} e_{j_k} \wedge \cdots \wedge \widehat{e_{i_l}} \wedge \cdots \wedge e_{i_{p+1}}.$$

On the other hand,

$$
hd(e_{i_1} \wedge \cdots \wedge e_{i_{p+1}}) = h\left(\sum_{l=1}^{p+1} (-1)^{l+1} f_{i_l} e_{i_1} \wedge \cdots \wedge \widehat{e}_{i_l} \wedge \cdots \wedge e_{i_{p+1}} \right) =
$$

$$
= \sum_{l=1}^{p+1} (-1)^{l+1} f_{i_l} (g_{i_l} e_{i_1} \wedge e_{i_1} \wedge \cdots \wedge \widehat{e}_{i_l} \wedge \cdots \wedge e_{i_{p+1}} +
$$

$$
+ \sum_{k=1}^{p+1} g_{j_k} e_{j_k} \wedge e_{i_1} \wedge \cdots \wedge \widehat{e}_{i_l} \wedge \cdots \wedge e_{i_{p+1}} =
$$

$$
= \sum_{l=1}^{p+1} (-1)^{l+1} f_{i_l} (-1)^{l-1} g_{i_l} e_{i_1} \wedge \cdots \wedge e_{i_{p+1}} +
$$

$$
+ \sum_{k=1}^{p+1} g_{j_k} e_{j_k} \wedge e_{i_1} \wedge \cdots \wedge \widehat{e}_{i_l} \wedge \cdots \wedge e_{i_{p+1}}.
$$

Adding up these two expressions we get the result desired.   □

2.8.5c.i Corollary *If the ideal of A generated by the elements f_i, where $i = 1, \ldots, r$, is equal to A, then the complexes $C_n^p(M)$ and $\overline{C(U, \mathcal{F})}$ are acyclic in dimensions ≥ 1.* (Indeed, then $\sum_{i=1}^{r} f_i g_i = 1$ for certain g_i, and hence h is a homotopy operator.)

2.8.5c.ii *Remark* In the general case, $C_n^{p+1}(M) = K^{p+1}(f^n, M)$ for $p \geq 0$. By definition, we have an exact sequence

$$
0 \longrightarrow H^0(f^n, M) \longrightarrow K^0(f^n, M) \xrightarrow{\ d\ } Z^1(f^n, M) \longrightarrow H^1(f^n, M) \longrightarrow 0.
$$

The limit $\varinjlim$ as $n \longrightarrow \infty$ gives

$$
0 \longrightarrow H^0((f), M) \longrightarrow K^0((f), M) \xrightarrow{\ d\ } Z^1((f), M) \longrightarrow H^1((f), M) \longrightarrow 0,
$$

and

$$
H^0(U, \mathcal{F}) = Z^1((f), M),
$$

$$
K^0((f), M) = K^0((f^n), M) = \operatorname{Hom}(A, M) = M,
$$

so we have the exact sequence

$$
0 \longrightarrow H^0((f), M) \longrightarrow M \xrightarrow{\ d\ } H^0(U, \mathcal{F}) \longrightarrow H^1((f), M) \longrightarrow 0.
$$

2.9 Cohomology of the Projective Space

Let A be a fixed ring, $\mathbb{P}_A^{r-1} = \operatorname{Proj} R$, where $R = A[T_1, \ldots, T_r]$ with the standard grading, $U_i = D_+(T_i)$, $U = (U_i)$.

In this section we compute the group $H^p(\mathbb{P}_A^{r-1}, \mathcal{O}(n))$ for any p, n, and r. This computation, due to Serre, is the basis of the proof (in the next section) of main results on cohomology of coherent sheaves on projective schemes.

Thanks to results presented in Sect. 2.8, we have

$$H^p(\mathbb{P}_A^{r-1}, \mathcal{O}(n)) = \check{H}^p(U, \mathcal{O}(n)).$$

Therefore, we can compute the cohomology of the Čech complex of the cover U.
Since

$$U_{i_0, \ldots, i_p} = D_+(T_{i_0} \cdots T_{i_p}) = \operatorname{Spec} R_{(T_{i_0} \cdots T_{i_p})},$$

we have, for the usual definition of the sheaf $\mathcal{O}(n)$, see Eq. (2.5) and Theorem 2.7.5c:

$$\Gamma(U_{i_0 \ldots i_p}, \mathcal{O}(n)) = \left\{ \frac{m(i_0, \ldots, i_p)}{(T_{i_0} \cdots T_{i_p})^k} \;\middle|\; k \in \mathbb{Z}, \; m \in R_{k(p+1)+n} \right\}.$$

This implies that

$$\bigoplus_{n \in \mathbb{Z}} \Gamma(U_{i_0 \ldots i_p}, \mathcal{O}(n)) = \{ s_{i_0 \ldots i_p} \mid s_{i_0 \ldots i_p} \in R_{(T_{i_0} \cdots T_{i_p})} \}.$$

This formula indicates that it is convenient to compute the direct sum of Čech complexes $\bigoplus_{n \in \mathbb{Z}} C^0(U, \mathcal{O}(n))$, and its cohomology, keeping track of the natural grading, and at the end separate the homogeneous components in the answer.

2.9.1 $C_k^p(U, \mathcal{O}(n))$

Fix $k \in \mathbb{Z}$. Denote by $C_k^p(U, \mathcal{O}(n))$ the subgroup of cochains whose components can be represented as $\dfrac{m(i_0, \ldots, i_p)}{(T_{i_0} \ldots T_{i_p})^k}$, where m is a form. As p varies, these groups form a complex $\bigoplus_{n \in \mathbb{Z}} C_k^p(U, \mathcal{O}(n))$; computing the action of its differential on the numerators of the components of cochains we easily obtain:

2.9.1a Lemma

(1) *The complex $\bigoplus_{n\in\mathbb{Z}} C_k^p(U, \mathcal{O}(n))$ with its grading is isomorphic to the Koszul complex $K^{p+1}(T_1^k, \ldots, T_r^k; R)$ with the grading in which the elements*

$$g \in \operatorname{Hom}(K_{p+1}(T^k), R) \quad \text{such that} \quad g(e_{i_0} \wedge \cdots \wedge e_{i_p}) \in R_{k(p+1)+n}$$

are homogeneous of degree n.

(2) *The map*

$$m(i_0, \ldots, i_p) \longmapsto T_{i_0} \cdots T_{i_p} \cdot m(i_0, \ldots, i_p)$$

defines homogeneous homomorphisms of graded complexes

$$\bigoplus_{n\in\mathbb{Z}} C_k^p(U, \mathcal{O}(n)) \longrightarrow \bigoplus_{n\in\mathbb{Z}} C_{k+1}^p(U, \mathcal{O}(n))$$

and

$$\bigoplus C^p(U, \mathcal{O}(n)) = \varinjlim_k \bigoplus C_k^p(U, \mathcal{O}(n))$$

with respect to this system of homomorphisms.

Observe that condition (1) uniquely determines on K^{p+1} the structure of a graded R-module.

Lemma 2.9.1a illustrates the necessity of studying the homology of the Koszul complex $K^p(T^k, R)$. The method of the chain homotopy operator is inapplicable because the elements $(T_1^k, \ldots, T_n^k)$ generate a non-trivial ideal; in fact, the Koszul complex is not acyclic in one of the dimensions.[8] Therefore, another approach is needed here.

We return, temporarily, to the notation of Sect. 2.8: let $f = (f_1, \ldots, f_r)$ be a collection of elements in a ring A. First, note the following duality.

2.9.2 Lemma *Define an A-homomorphism*

$$\varphi: K_{r-p}(f, A) \longrightarrow K^p(f, A)$$

by setting

$$\varphi(e_{i_1} \wedge \cdots \wedge e_{i_{r-p}})(e_{j_1} \wedge \cdots \wedge e_{j_p}) =$$

$$= \begin{cases} 0, & \text{if } (i) \cap (j) \neq \emptyset, \\ \varepsilon(i_1, \ldots, i_{r-p}; j_1, \ldots, j_p), & \text{if } (i) \cap (j) = \emptyset. \end{cases}$$

Then φ is a complex isomorphism up to a sign of the differentials.

---

[8] **Exercise.** Determine in which one.

Proof It suffices to find how φ commutes with the differentials. We have:

$$\varphi(d(e_{i_1} \wedge \cdots \wedge e_{i_{r-p}}))(e_{j_1} \wedge \cdots \wedge e_{j_{p+1}}) =$$

$$= \varphi\left(\sum_{k=1}^{p} (-1)^{k+1} f_{i_k} e_{i_1} \wedge \cdots \wedge \widehat{e}_{i_k} \wedge \cdots \wedge e_{i_{r-p}} \right)(e_{j_1} \wedge \cdots \wedge e_{j_{p+1}}) =$$

$$= \begin{cases} 0, & \text{if } |(i) \cap (j)| > 1, \\ (-1)^{k+1+\sigma} f_{i_k}, & \text{if } i_k = (i) \cap (j), \end{cases}$$

where

$$(-1)^{\sigma} = \varepsilon(i_1, \ldots, \widehat{i_k}, \ldots i_{r-p}; j_1, \ldots, j_{p+1}).$$

On the other hand,

$$\varphi(e_{i_1} \wedge \cdots \wedge e_{i_{r-p}})(d(e_{j_1} \wedge \cdots \wedge e_{j_{p+1}})) =$$

$$= \varphi(e_{i_1} \wedge \cdots \wedge e_{i_{r-p}})\left(\sum_{l=1}^{p+1} (-1)^{l+1} e_{j_1} \wedge \cdots \wedge e_{j_l} \wedge \cdots \wedge e_{j_{p+1}} \right) =$$

$$= \begin{cases} 0, & \text{if } |(i) \cap (j)| > 1, \\ (-1)^{l+1+\tau} f_{j_l}, & \text{if } j_l = (i) \cap (j), \end{cases}$$

where

$$(-1)^{\tau} = \varepsilon(i_1, \ldots, i_{r-p}; j_1, \ldots, \widehat{j_l}, \ldots, j_{p+1}).$$

Comparing these answers we see how φ commutes with the differentials. $\square$

Next, we use Lemma 2.8.5a, which implies that

$$K_0(f_1, \ldots, f_r) = K_0(f_1, \ldots, f_{r-1}) \otimes K_0(f_r).$$

The following result justifies the computation of the homology of K_0 by induction on r.

For any A-module M and element $f \in A$, let ${}_f M$ denote the kernel of the homomorphism $M \xrightarrow{f} M$ of left multiplication by f.

2.9.3 Lemma *Let L be a chain complex of A-modules, and let $i \geq 1$. Then there is an exact sequence of A-modules*

$$0 \longrightarrow H_i(L)/f \cdot H_i(L) \longrightarrow H_i(L \otimes K(f)) \longrightarrow {}_f H_{i-1}(L) \longrightarrow 0.$$

Proof The complex $K(f)$ is of the form

$$0 \longrightarrow Ae_1 \xrightarrow{\ d\ } Ae_0 \longrightarrow 0, \quad \text{where } de_1 = fe_0.$$

Therefore, for $i \geq 1$ we have

$$(L \otimes K(f))_i = L_i e_0 \oplus L_{i-1} e_1$$

and

$$d(l_i e_0 + l_{i-1} e_1) = (dl_i + (-1)^{i-1} f l_{i-1}) e_0 + dl_{i-1} e_1.$$

This yields a commutative diagram with exact rows

$$
\begin{array}{ccccccccc}
0 & \longrightarrow & L_i & \longrightarrow & (L \otimes K(f))_i & \longrightarrow & L_{i-1} & \longrightarrow & 0 \\
 & & \downarrow{\scriptstyle d} & & \downarrow{\scriptstyle d} & & \downarrow{\scriptstyle d} & & \\
0 & \longrightarrow & L_{i-1} & \longrightarrow & (L \otimes K(f))_{i-1} & \longrightarrow & L_{i-2} & \longrightarrow & 0
\end{array}
$$

All these diagrams can be united into a sequence of complexes that is exact in dimensions ≥ 1:

$$0 \longrightarrow L \longrightarrow (L \otimes K(f)) \longrightarrow L[-1] \longrightarrow 0, \tag{2.13}$$

where $L[-1]_i := L_{i-1}$ (the complex L, shifted by 1 to the right). In its turn, this sequence leads to an exact sequence of homology groups

$$
\cdots \longrightarrow H_{i+1}(L[-1]) \xrightarrow{\ \delta\ } H_i(L) \longrightarrow H_i(L \otimes K(f)) \longrightarrow H_i(L[-1]) \xrightarrow{\ \delta\ } \cdots
$$
$$
\qquad\quad \| \qquad\qquad\qquad\qquad\qquad\qquad\qquad\qquad\qquad\quad \|
$$
$$
\qquad H_i(L) \qquad\qquad\qquad\qquad\qquad\qquad\qquad\qquad\quad H_{i-1}(L)
$$

Let us show that the diagram

$$
\begin{array}{ccc}
H_{i+1}(L[-1]) & \xrightarrow{\ \delta\ } & H_i(L) \\
\| & \nearrow & \\
H_i(L) & {\scriptstyle [-1]^i f} &
\end{array}
$$

is commutative. This follows from calculations: if $z \in Z_{i+1}(L[-1])$, then

$$d(z \otimes e_1) = dz \otimes e_1 + (-1)^i z \otimes f e_0 = (-1)^i z f \otimes e_0,$$
$$\delta(\text{class}(z)) = (-1)^i \, \text{class}(fz).$$

Thus, the following sequence is exact:

$$0 \longrightarrow H_i(L)/fH_i(L) \longrightarrow H_i(L \otimes K(f)) \longrightarrow {}_fH_{i-1}(L) \longrightarrow 0.$$

2.9.3a Corollary *If $H_i(L) = 0$ for $i \geq 1$, then $H_i(L \otimes K(f)) = 0$ for $i \geq 2$.*

Applying Lemma 2.9.3 and Corollary 2.9.3a to the Koszul complex one obtains the following result:

2.9.4 Proposition *Let $f_1, \ldots, f_r \in A$ be a sequence of elements such that f_i is not a zero divisor in $A/(f_1, \ldots, f_{i-1})$ for all $i \geq 1$. Then $H_i(K(f)) = 0$ for $i \geq 1$ and $f = (f_1, \ldots, f_r)$. Moreover, we always have*

$$H_0(K(f)) = A/(f_1, \ldots, f_r)A.$$

Proof easily follows by induction on r. Indeed, if the statement holds for the sequences of length $i - 1$, then Corollary 2.9.3a immediately shows that acyclicity holds in dimensions ≥ 2 for any sequence of length i, whereas the exact sequence (2.13) shows that acyclicity holds in dimension 1 as well. The claim on H_0 is obvious from the definition. $\square$

2.9.5 Regular Sequences

Any sequence $f_1, \ldots, f_r \in A$ satisfying the conditions in Proposition 2.9.4 is said to be *regular*.

The result of the above subsection shows that if the elements $f_1, \ldots, f_r$ form a regular sequence, then the Koszul complex is a free resolution of the A-module $A/(f_1, \ldots, f_r)$.

In several important cases the converse of Proposition 2.9.4 holds:

2.9.6 Proposition *Let A be a local Noetherian ring, $\mathfrak{p}$ its maximal ideal, and $f_1, \ldots, f_r \in \mathfrak{p}$. If $H_i(K(f_1, \ldots, f_r)) = 0$ for $i \geq 1$ (or even if only $H_1(K(f_1, \ldots, f_r)) = 0$), then $f_1, \ldots, f_r$ is a regular sequence.*

Proof For $r = 1$ the proposition obviously holds for any A and $f_1 \in A$. Suppose the statement is proved for $r = n - 1$; let $f_1, \ldots, f_n \in \mathfrak{p}$ and $H_1(K(f_1, \ldots, f_n)) = 0$. Applying Lemma 2.9.3 for $f = f_n$, and $i = 1$, we get the exact sequence

$$0 \longrightarrow H_1(K(f_1, \ldots, f_{n-1}))/fH_1(K(f_1, \ldots, f_{n-1})) \longrightarrow$$
$$\longrightarrow H_1(K(f_1, \ldots, f_n)) \longrightarrow {}_fH_1(K(f_1, \ldots, f_{n-1})) \longrightarrow 0,$$

which implies that

$$H_1(K(f_1,\ldots,f_{n-1}))/fH_1(K(f_1,\ldots,f_{n-1})) = 0, \qquad (2.14)$$

$$_fH_0(K(f_1,\ldots,f_{n-1})) = 0. \qquad (2.15)$$

Since under the conditions of the proposition $H_1(K(f_1,\ldots,f_{n-1}))$ is a Noetherian A-module, it follows from (2.14) and Nakayama's lemma that

$$H_1(K(f_1,\ldots,f_{n-1})) = 0,$$

so $f_1,\ldots,f_{n-1}$ is a regular sequence by induction hypothesis. Further, since

$$H_0(K(f_1,\ldots,f_{n-1})) = A/(f_1,\ldots,f_{n-1}),$$

it follows that (2.15) implies that f_n is not a zero divisor in $A/(f_1,\ldots,f_{n-1})$. Hence, $f_1,\ldots,f_n$ is a regular sequence. □

2.9.7 Corollary *If $(f_1,\ldots,f_n)$ is a regular sequence of elements of the maximal ideal of a Noetherian local ring, then the sequence obtained from it by any permutation is also regular.*

Proof It suffices to observe that the Koszul complexes corresponding to two sequences that differ only by a permutation are isomorphic. □

We are now able to formulate the main result on the cohomology of the projective space.

2.9.8 Theorem

(a) $H^p(\mathbb{P}_A^{r-1}, \mathcal{O}(n)) = 0$ *for* $p \neq 0, r-1$.

(b) $H^0(\mathbb{P}_A^{r-1}, \mathcal{O}(n)) = 0$ *for* $n < 0$ *and is a free A-module of rank* $\binom{n+r-1}{r-1}$ *for* $n \geq 0$.

(c) $H^{r-1}(\mathbb{P}_A^{r-1}, \mathcal{O}(n)) = 0$ *for* $n \geq -r+1$ *and is a free A-module of rank* $\binom{-n-1}{r-1}$ *for* $n \leq -r$.

On the (p,n)-plane, mark the points where $H^p(\mathbb{P}_A^{r-1}, \mathcal{O}(n)) \neq 0$. Obviously, the resulting diagram is centrally symmetric, see Fig. 2.5, and this symmetry $(p,n) \longleftrightarrow (r-1-p, -r-n)$ preserves the rank of the cohomology A-modules. In a deeper theory, this symmetry is explained by the duality theorem for cohomology of coherent sheaves.

Proof We only have to keep track of the homomorphisms connecting $H^p(\mathbb{P}_A^{r-1}, \mathcal{O}(n))$ with the corresponding Koszul complexes.

The fact that the Koszul complex $K_0(T^k, R)$ is acyclic (Proposition 2.9.4) and the duality described in Lemma 2.9.2 immediately imply statement (a) if we realize that the role of the ring A is played now by $R = A[T_1,\ldots,T_r]$, and the role of f_i is played by T_i for each i.

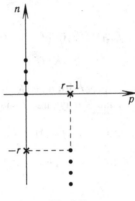

Fig. 2.5

A somewhat more tedious calculation with the grading and the explicit form of homomorphisms $C_k \longrightarrow C_{k+1}$ (see Lemma 2.9.1a) taken into account allows one to establish statements (b) and (c). We leave this calculation to the reader. $\qquad\square$

2.10   Serre's Theorem

2.10.1 Theorem (Serre) *Let R be a graded ring: $R = \bigoplus_{n=0}^{\infty} R_n$, where $R_0 = A$ is a Noetherian ring and R is generated by the Noetherian A-module R_1. Let $X = \operatorname{Proj} R$ and $\mathcal{F}$ be a coherent sheaf on X. Then the following statements hold:*

(a) $H^q(X, \mathcal{F}) = 0$ *if $q + 1$ is greater than the number of generators of the A-module R_1.*

(b) $H^q(X, \mathcal{F})$ *is a Noetherian A-module for any q.*

(c) $H^q(X, \mathcal{F}(n)) = 0$ *for $q \geq 1$ and $n \geq n_0(\mathcal{F})$, where $n_0(\mathcal{F})$ is a number depending on $\mathcal{F}$.*

Observe that this theorem allows one to introduce important invariants of the scheme X. For example, if A is a field and $\mathcal{F}$ is the structure sheaf, then the cohomology spaces $H^q(X, \mathcal{O}_X)$ are specified by their dimensions, which are determined by X.

Proof First of all, we reduce Serre's theorem to the case where $X = \mathbb{P}_A^r$, where $r + 1$ is the cardinality of a set of generators of the A-module R_1. Consider the ring $A[T_1, \ldots, T_{r+1}]$, whose projective spectrum is $\mathbb{P}_A^r$, and construct a   homogeneous epimorphism of graded rings $A[T_1, \ldots, T_{r+1}] \longrightarrow R$ by sending T_i to the ith generator of R_1. This epimorphism induces a closed embedding of spectra $j : \operatorname{Proj} R \longrightarrow \mathbb{P}_A^r$, which in turn enables us to continue the sheaf $\mathcal{F}$ on $\operatorname{Proj} R$, figuring in the statement of the theorem, to the sheaf $j_*(\mathcal{F})$ on $\mathbb{P}_A^r$. The sheaf $j_*(\mathcal{F})$ is defined by the rule

$$\Gamma(U, j_*(\mathcal{F})) = \Gamma(U \cap j(X), \mathcal{F}) \text{ for any open set } U \subset \mathbb{P}^r_A.$$

The following properties of the sheaf continuation operation are pretty obvious. First,

$$H^q(X, \mathcal{F}) = H^q(\mathbb{P}^r_A, j_*(\mathcal{F})) \text{ for any } q,$$

which is easy to see on the level of Čech complexes: in the present case they are just isomorphic as graded modules with differentials. (As opens in $\mathbb{P}^r_A$ we take, as always, the sets $D_+(T_i)$; as opens in $\operatorname{Proj} R$ we take, in order to establish the aforementioned isomorphism of complexes, the sets $D_+(t_i)$, where t_i is the generator of R_1 corresponding to T_i.) We may also use of the fact that $\mathcal{F} \longrightarrow j_*(\mathcal{F})$ is a fully faithful functor.[9]

Second, we have $j_*(\mathcal{F}(n)) = j_*(\mathcal{F})(n)$, where

$$\mathcal{F}(n) := \mathcal{F} \otimes \mathcal{O}_X(n).$$

Now, item (a) of Serre's theorem immediately follows from the antisymmetry of Čech cochains and the fact that the Čech cohomology for the cover $(D_+(T_i))$ coincides with the usual cohomology.

To prove items (b) and (c) of Serre's theorem we use the following technical result.

2.10.1a Lemma *For any coherent sheaf $\mathcal{F}$ on $\mathbb{P}^r_A$, there exists an integer m such that for some natural p we have the exact sequence*

$$\mathcal{O}^p_{\mathbb{P}^r_A} \longrightarrow \mathcal{F}(m) \longrightarrow 0.$$

From this lemma the needed properties (b) and (c) are established by a simple descending induction on q.

Let us define the coherent sheaf $\mathcal{E}$ by means of the exact sequence

$$0 \longrightarrow \mathcal{E} \longrightarrow \mathcal{O}^p_{\mathbb{P}^r_A} \longrightarrow \mathcal{F}(m) \longrightarrow 0.$$

Next, we tensor this sequence by $\mathcal{O}(n)$, where $n \in \mathbb{Z}$:

$$0 \longrightarrow \mathcal{E}(n) \longrightarrow \mathcal{O}^p_{\mathbb{P}^r_A}(n) \longrightarrow \mathcal{F}(m+n) \longrightarrow 0.$$

This yields the exact cohomology sequence:

---

[9] A given functor is *faithful* (resp. *full*) if it is injective (resp. surjective) when restricted to each set of morphisms that have a given source and target.

$$\cdots \longrightarrow H^q(\mathbb{P}^r_A, \mathcal{O}(n)) \longrightarrow H^q(\mathbb{P}^r_A, \mathcal{F}(m+n)) \longrightarrow H^{q+1}(\mathbb{P}^r_A, \mathcal{E}(n)) \longrightarrow \cdots$$

$$(2.16)$$

For $q = r + 1$, item (a) shows that properties (b) and (c) hold trivially. The induction hypothesis: suppose items (b) and (c) of Serre's theorem hold for the cohomology of all coherent sheaves in dimension $q + 1$.

In the exact cohomology sequence (2.16), the A-module $H^{q+1}(\mathbb{P}^r_A, \mathcal{E}(n))$ is Noetherian and vanishes for $n \geq n_0$ by the induction hypothesis; for $H^q(\mathbb{P}^r_A, \mathcal{O}(n))$ the same is true thanks to Theorem 2.9.8. This yields what is needed for $H^q(\mathbb{P}^r_A, \mathcal{F}(m+n))$. $\square$

Proof of Lemma 2.10.1a We may assume that $\mathcal{F}$ is a coherent sheaf on $\mathbb{P}^r_A$. For the standard cover by the opens $U_i := D_+(T_i)$, by identifying $\mathcal{F}|_{U_i}$ with $\mathcal{F}(m)|_{U_i}$, we see that $\mathcal{F}(m)$ is glued from the $\mathcal{F}|_{D_+(T_i)}$'s by means of the cocycle

$$\left(\frac{T_i}{T_j}\right)^m : (\mathcal{F}|_{U_i})|_{U_i \cap U_j} \xrightarrow{\simeq} (\mathcal{F}|_{U_j})|_{U_j \cap U_i}.$$

Each section $s \in \Gamma(\mathbb{P}^r_A, \mathcal{F}(m))$ is then given by its "components" $s \in \Gamma(U_i, \mathcal{F})$: there is an embedding

$$\Gamma(\mathbb{P}^r_A, \mathcal{F}(m)) \hookrightarrow \prod \Gamma(U_i, \mathcal{F}),$$

$$s \longmapsto (\ldots, s_i, \ldots), \text{ where } s_i \cdot \left(\frac{T_i}{T_j}\right)^m = s_j.$$

To construct the epimorphism $\mathcal{O}^p \longrightarrow \mathcal{F}(m)$, it suffices to establish that one can construct a finite number of global sections of the sheaf $\mathcal{F}(m)$ whose restrictions to U_i generate the $\Gamma(U_i, \mathcal{O}_{\mathbb{P}^r_A})$-module $\Gamma(U_i, \mathcal{F})$.

To this end, it suffices to establish, for any section $s \in \Gamma(U_i, \mathcal{F})$ and all sufficiently large m, that s is a component of a sheaf $\mathcal{F}(m)$, i.e., to show that one can "extend s". Once we have extended in this way (finitely many) generators of all modules $\Gamma(U_i, \mathcal{F})$ and have selected a common m sufficiently large, the result will follow.

Extending a Section $s_0 \in \Gamma(U_0, \mathcal{F})$ After it is multiplied by $\left(\frac{T_0}{T_i}\right)^p$ for a certain p, the section $s_0|_{U_0 \cap U_i}$ can be extended to U_i:

$$s_0 \left(\frac{T_0}{T_i}\right)^p = s_i'|_{U_0 \cap U_i}, \text{ where } s_i' \in \Gamma(U_i, \mathcal{F});$$

this p is common for all sets U_i, and arbitrarily large.

On $U_0 \cap U_i \cap U_j$, we have:

$$s_i' \left(\frac{T_i}{T_j}\right)^p \Big|_{U_0 \cap U_i \cap U_j} = s_j'|_{U_0 \cap U_i \cap U_j}.$$

Therefore, for q sufficiently large (and also common for all i), we have

$$\left(s_i'\left(\frac{T_i}{T_j}\right)^p - s_j'\right)\left(\frac{T_0}{T_j}\right)^q = 0. \tag{2.17}$$

Now set

$$s_i'' := s_i'\left(\frac{T_0}{T_i}\right)^q.$$

Clearly,

$$s_0'' = s_0.$$

Moreover, s_0'' figures as a component in the section $(\ldots, s_i'', \ldots)$ of the sheaf $\mathcal{F}(p+q)$. Indeed (see (2.17)):

$$s_i''\left(\frac{T_i}{T_j}\right)^{p+q} = s_j'' \iff s_i'\left(\frac{T_0}{T_i}\right)^q\left(\frac{T_i}{T_j}\right)^{p+q} = s_j'\left(\frac{T_0}{T_j}\right)^q. \qquad \square$$

2.10.2 Comments to Serre's Theorem

(1) The role of the coherence of $\mathcal{F}$: without it property (b) fails.

Statement (b) is non-trivial: the modules C^q in the Čech complex are not of finite type: Noetherian property appears only after passage to cohomology.

(2) The nature of the proof: there are "plenty" of sheaves $\mathcal{O}_{\mathbb{P}^r}(n)$ with known cohomology; everything can be reduced to them with the help of an exact cohomology sequence. Lemma 2.10.1a explains the meaning of the term "plenty".

Compare with the following algebraic fact: every module is the image of a free module; here, however, the situation is global, and "twisting" by means of $\mathcal{O}_{\mathbb{P}^r}(1)$ is essential.

(3) What is the value $n_0(\mathcal{F})$, starting with which (c) holds? The question is difficult; for the sheaves of ideals we have, nevertheless, a "crude" answer: if the Hilbert polynomial h is known, then $n_0(\mathcal{F})$ is a universal polynomial in the coefficients of h. Generally, the numbers $n_0(\mathcal{F})$ depend on $\mathcal{F}$; for example, if $\mathcal{F} = \mathcal{O}_{\mathbb{P}^r}(-N)$, then $H^r(\mathbb{P}_A^r, \mathcal{F}(n)) = 0$ only for $n \geq N - r$.

2.11   Sheaves on Proj R and Graded Modules

2.11.1   A Basic Fact of the Theory of Quasi-coherent Sheaves over Affine Schemes $X = \operatorname{Spec} A$

The fact we have in mind is stated as follows:

The functor $\mathcal{F} \longrightarrow \Gamma(X, \mathcal{F})$ establishes an equivalence of the category of sheaves with the category of A-modules.

The purpose of this section is to show the existence of an analogous correspondence between the category of quasi-coherent sheaves on Proj R and the category of *graded* R-modules.

This correspondence is, however, not as simple and straightforward as in the affine case; in particular, graded modules that differ in only finitely many components lead to isomorphic sheaves. For a precise formulation, see Sect. 2.11.6.

2.11.2   Let $X = \operatorname{Proj} R$

Let $X = \operatorname{Proj} R$, where the ring R is generated by the set R_1 over R_0. Let $\mathcal{F}$ be a quasi-coherent sheaf over X and set

$$\Gamma_*(X, \mathcal{F}) := \bigoplus_{n=0}^{\infty} \Gamma(X, \mathcal{F}(n)).$$

If $\mathcal{F} = \mathcal{O}_X$, then on $\Gamma_*(X, \mathcal{O}_X)$ there exists a natural structure of a graded ring with the multiplication induced by the homomorphisms

$$\mathcal{O}_X(n) \otimes \mathcal{O}_X(m) \longrightarrow \mathcal{O}_X(m + n).$$

More generally, the multiplications $\mathcal{O}_X(n) \otimes \mathcal{F}(m) \longrightarrow \mathcal{F}_X(m + n)$ determine a structure of graded $\Gamma_*(X, \mathcal{O}_X)$-module on $\Gamma_*(X, \mathcal{F})$.

Now recall that, for any n, there are defined group homomorphisms

$$\alpha_n \colon R_n \longrightarrow \Gamma(X, \mathcal{O}_X(n)).$$

(see Theorem 2.7.4). The definition makes clear that these homomorphisms are compatible with multiplication and hence yield a homogeneous homomorphism of graded rings

$$\alpha \colon R \longrightarrow \Gamma_*(X, \mathcal{O}_X).$$

Therefore, *there is a canonical structure of graded R-module on $\Gamma_*(X, \mathcal{F})$.*

2.11.3 Conversely, let M be a Graded R-module

From M we construct a quasi-coherent sheaf $\widetilde{M}$ on X by setting

$$\Gamma(D_+(f), \widetilde{M}) = M_{(f)} \text{ for any } f \in R_1$$

and defining the restriction homomorphisms as in Sect. 2.4 for $M = R$.

The construction of the homomorphism $\alpha: M \longrightarrow \Gamma_*(X, \widetilde{M})$ is easy to be translated, component-wise, to this case:

$$\alpha_n: M_n \longrightarrow \Gamma_*(X, \widetilde{M}(n)), \quad \alpha_n(m)|_{D_+(f)} = m/f^n \in M_{(f)}.$$

2.11.3a Exercise Verify the natural compatibilities.

2.11.4 Proposition *Any quasi-coherent sheaf $\mathcal{F}$ on X is isomorphic to a sheaf of the form $\widetilde{M}$, where $M = \Gamma_*(X, \mathcal{F})$.*

Proof First of all, let us construct an isomorphism

$$\beta: \widetilde{M} = \widetilde{\Gamma_*(X, \mathcal{F})} \longrightarrow \mathcal{F}.$$

It suffices to construct β for the sections of the corresponding sheaves over opens $D_+(f)$. We have

$$\Gamma(D_+(f), \widetilde{M}) = \widetilde{M}_{(f)} = \left[\bigoplus_{n=0}^{\infty} \Gamma(X, \mathcal{F}(n)) \right]_{(f)} = \left\{ \frac{m}{f^n} \mid m \in \Gamma(X, \mathcal{F}(n)) \right\}.$$

Set

$$\beta\left(\frac{m}{f^n}\right) = \alpha(m)|_{D_+(f)} \cdot \alpha(f)^{-n}|_{D_+(f)} \in \Gamma(D_+(f)).$$

2.11.4a Exercise Verify that β is well defined.

It only remains to establish that β is an isomorphism. In the same way as for the structure sheaf, one can prove that $\operatorname{Ker} \alpha_n = 0$ for n sufficiently large. This implies that β is a monomorphism; indeed,

$$\alpha(m)\alpha(f)^{-n} = 0 \Longrightarrow \alpha(f^e m) = 0, \text{ for } e \geq e_0 \Longrightarrow m/f^n = 0.$$

Lemma 2.10.1a immediately implies that β is also an epimorphism: indeed, to extend the section $s \in \Gamma(D_+(f), \mathcal{F})$ to a section of $\mathcal{F}(n)$ over $D_+(f)$ means precisely to represent s as the image of an element m/f^n.   □

The following result, analogous to a result obtained in Sects. 2.11.2 and 2.11.3, is deeper, but concerns the homomorphism α.

2.11.5 Theorem *Let the ring R satisfy the conditions of Serre's theorem 2.10.1, M a graded Noetherian R-module, and $\mathcal{F} = \widetilde{M}$. Then the sheaf $\mathcal{F}$ is coherent and the map*

$$\alpha_n : M_n \longrightarrow \Gamma(X, \mathcal{F}(n))$$

is an isomorphism for sufficiently large n.

Proof First of all, as in Serre's theorem, it is easy to see that it suffices to carry out the proof for the case $X = \mathbb{P}_A^r$, which is how we proceed in what follows.

There exists an exact sequence

$$L \xrightarrow{f} L' \longrightarrow M \longrightarrow 0,$$

where L and L' are free graded R-modules, i.e., direct sums of R-modules $R(n)$.

This gives an exact sequence of sheaves

$$\widetilde{L}(n) \xrightarrow{\widetilde{f}(n)} \widetilde{L}'(n) \longrightarrow \widetilde{M}(n) \longrightarrow 0,$$

which immediately implies that the sheaf $\mathcal{F}$ is coherent, and with the help of which one constructs the exact cohomology sequence in the bottom row of the commutative diagram

$$
\begin{array}{ccccccc}
L_n & \longrightarrow & L'_n & \longrightarrow & M_n & \longrightarrow & 0 \\
\downarrow{\scriptstyle\alpha} & & \downarrow{\scriptstyle\alpha} & & \downarrow{\scriptstyle\alpha} & & \searrow \\
\Gamma(X, \widetilde{L}(n)) & \longrightarrow & \Gamma(X, \widetilde{L}'(n)) & \longrightarrow & \Gamma(X, \mathcal{F}(n)) & \longrightarrow & H^1(X, f\widetilde{(L)}(n))
\end{array}
$$

By Serre's theorem, $H^1(X, f\widetilde{(L)}(n)) = 0$ for $n \geq n_0$, and because $X = \mathbb{P}_A^r$, Lemma 2.7.5d shows that the first two vertical arrows are isomorphisms. Hence, $\alpha : M_n \longrightarrow \Gamma(X, \mathcal{F}(n))$ is also an isomorphism.   □

2.11.6   Basic Theorem on the Correspondence $\mathcal{F} \longmapsto \Gamma_*(X, \mathcal{F})$

We formulate it without proof: a good deal of it is already verified; to prove the rest does not require any new ideas.

Let R be a graded ring satisfying the conditions of Serre's theorem. Denote by GM_R the *category* (of graded modules) defined as follows:

the objects of GM_R are graded Noetherian R-modules;
the morphisms in GM_R: for any two R-modules $M = \bigoplus M_i$ and $N = \bigoplus N_i$ we set

$$\mathrm{Hom}(M, N) = \varinjlim_{i_0} \mathrm{Hom}_R \left(\bigoplus_{i \geq i_0} M_i, \ \bigoplus_{i \geq i_0} N_i \right),$$

where $\varinjlim$ is applied to the group Hom_R of homogeneous R-homomorphisms of graded $\overrightarrow{R}$-modules.

Informally speaking, a morphism in the category GM_R is represented by a homomorphism of modules $M \longrightarrow N$ that is defined only starting with components of sufficiently high degree; two homomorphisms coinciding in all sufficiently high degrees define the same morphism.[10]

Modules M and N, isomorphic as objects of GM_R, are said to be *TN-isomorphic*.

2.11.6a Theorem *The functor $\mathcal{F} \longmapsto \Gamma_*(X, \mathcal{F})$ establishes an equivalence of the category of coherent sheaves on $X = \mathrm{Proj}\,R$ with GM_R; its inverse is the functor*

$$M \longmapsto \widetilde{M}.$$

2.12 Applications to the Theory of Hilbert Polynomials

The results of the preceding section allow one to give a "geometric" definition of the Hilbert polynomial and prove the invariance of a number of numerical characteristics introduced in Sect. 2.5.

In what follows, R is a ring satisfying the conditions of Serre's theorem and such that, in addition, $R_0 = k$ is a *field* (this is needed to count dimensions of homogeneous components; all the ensuing results can be easily generalized to the case where R_0 is an Artinian ring by considering lengths of modules instead of dimensions of linear spaces).

---

[10]The following two conditions on a graded R-module M are usually considered:
  (TF) There exists an n such that $\bigoplus_{k \geq n} M_k$ is a finitely generated R-module.
  (TN) There exists an n such that $\bigoplus_{k \geq n} M_k = 0$ for $k \geq n$.
  A graded R-module module homomorphism f is called *TN-injective* (resp. *TN-surjective*, resp. *TN-bijective*) if its kernel (resp. cokernel, resp. both) satisfies (TN).

2.12.1 Theorem *Let $X = \operatorname{Proj} R$, M a Noetherian R-module, $\mathcal{F} = \widetilde{M}$, and $h_M(n)$ the Hilbert polynomial of M. Then for all $n \in \mathbb{Z}$, we have*

$$h_M(n) = \sum_{i=0}^{\infty} (-1)^i \dim H^i(X, \mathcal{F}(n)) \overset{\text{def}}{=} \chi(\mathcal{F}(n)).$$

The polynomial $\chi(\mathcal{F})$ is called the *Euler characteristic* of the sheaf $\mathcal{F}$.

Proof Serre's theorem and Theorem 2.11.4 imply that

$$h_M(n) = \dim M_n = \dim H^0(X, \mathcal{F}(n)) = \chi(\mathcal{F}(n)) \text{ for } n \geq n_0.$$

Therefore, the theorem is obtained if we establish that $\chi(\mathcal{F}(n))$ can be represented as a polynomial in n *for all* n. The idea is the same as in the proof of existence of the Hilbert polynomial. It is based on the following lemma.

2.12.1a Lemma *Let*

$$0 \longrightarrow \mathcal{F}_1 \longrightarrow \mathcal{F} \longrightarrow \mathcal{F}_2 \longrightarrow 0$$

be an exact sequence of coherent sheaves on X. Then

$$\chi(\mathcal{F}) = \chi(\mathcal{F}_1) + \chi(\mathcal{F}_2).$$

Proof Let $r = \dim R_1$. Then $H^i(X, \mathcal{F}) = 0$ for $i \geq r$. The exact cohomology sequence

$$0 \longrightarrow H^0(X, \mathcal{F}_1) \longrightarrow H^0(X, \mathcal{F}) \longrightarrow \ldots \longrightarrow H^{r-1}(X, \mathcal{F}) \longrightarrow H^{r-1}(X, \mathcal{F}_2) \longrightarrow 0$$

implies that the alternating sum of dimensions of these cohomology spaces vanishes. This establishes the lemma.   □

2.12.1b Corollary *For any exact sequence*

$$0 \longrightarrow \mathcal{F}_0 \longrightarrow \cdots \longrightarrow \mathcal{F}_k \longrightarrow 0$$

of coherent sheaves on X we have

$$\sum_{i=0}^{k} (-1)^i \chi(\mathcal{F}_i) = 0.$$

Now, we use induction on r to prove that $\chi(\mathcal{F}(n))$ is a polynomial in n. A standard reduction allows us to assume that $X = \mathbb{P}_k^{r-1} = \operatorname{Proj} k[T_1, \ldots, T_r]$. Let $M \overset{T_r}{\longrightarrow} M(1)$ be the homomorphism of multiplication by T_r. This homomorphism preserves the grading: $M(1)_i = M_{i+1}$. If K and C are the kernel and the cokernel of this homomorphism, then the sequence

$$0 \longrightarrow K \longrightarrow M \xrightarrow{T_r} M(1) \longrightarrow C \longrightarrow 0$$

is exact. We derive from it an exact sequence of sheaves

$$0 \longrightarrow \widetilde{K} \longrightarrow \mathcal{F} \longrightarrow \mathcal{F}(1) \longrightarrow \widetilde{C} \longrightarrow 0$$

as well as an exact sequence of twisted sheaves

$$0 \longrightarrow \widetilde{K}(n) \longrightarrow \mathcal{F}(n) \longrightarrow \mathcal{F}(n+1) \longrightarrow \widetilde{C}(n) \longrightarrow 0.$$

By Corollary 2.12.1b,

$$\chi(\mathcal{F}(n)) - \chi(\mathcal{F}(n+1)) = \chi(\widetilde{K}(n)) - \chi(\widetilde{C}(n)).$$

The sheaves $\widetilde{K}$ and $\widetilde{C}$ correspond to $k[T_1, \ldots, T_{r-1}]$-modules, i.e., are "concentrated" on $\mathbb{P}_k^{r-1}$. This allows one to make the induction step: for $r = 0$, the statement is trivial. $\square$

2.12.1c *Comment*

Theorem 2.12.1 shows that the Hilbert polynomial is invariantly determined by the triple of geometric objects $(X, \mathcal{L}, \mathcal{F})$, where $X = \operatorname{Proj} R$, $\mathcal{L} = \mathcal{O}_X(1)$, and $\mathcal{F}$ is a coherent sheaf on X. Let us show that in fact the *degree* of this polynomial and its *constant term* depend only on $(X, \mathcal{F})$, and not on $\mathcal{L}$.

In particular, the *dimension* $\dim X$ and the *characteristic* $\chi(X)$ introduced above do not depend on the representation of the space X as $\operatorname{Proj} R$.

The claim concerning the constant term is obvious:

$$h_M(0) = \chi(\mathcal{F}) = \sum_{i=0}^{\infty} (-1)^i \dim H^i(X, \mathcal{F}).$$

2.12.2 Theorem *Let* $X = \operatorname{Proj} R$, $\mathcal{L} = \mathcal{O}_X(1)$, *and* $\mathcal{F}$ *is a coherent sheaf on* X. *Set*

$$\dim \mathcal{F} = \deg \chi(\mathcal{F}(n)).$$

Then $\dim \mathcal{F}$ *does not depend on the choice of invertible sheaf* $\mathcal{L}$.

Proof The formulation of the theorem assumes that the sheaf $\mathcal{L}$ is *very ample*, i.e., is of the form $\mathcal{O}(1)$ for a representation of X in the form $\operatorname{Proj} R$. We have no means of characterizing such sheaves apart from Serre's theorem, which is what we will use.

Let $\mathcal{L}_1$ and $\mathcal{L}_2$ be two very ample sheaves on X. For any $n \geq n_0$, we have

$$h_i(n) = \dim H^0(X, \mathcal{F} \otimes \mathcal{L}_i^n), \quad \text{where } i = 1, 2.$$

The sheaf $\mathcal{M} = \mathcal{L}_1^{-1} \otimes \mathcal{L}_2^N$ for N sufficiently large is generated by its sections, as follows from Lemma 2.10.1a.

Since $(\mathcal{F} \otimes \mathcal{L}_1^n) \otimes \mathcal{M}^n = \mathcal{F} \otimes \mathcal{L}_2^{nN}$, we have an isomorphism of groups of global sections

$$H^0(X, \mathcal{F} \otimes \mathcal{L}_1^n \otimes \mathcal{M}^n) \simeq H^0(X, \mathcal{F} \otimes \mathcal{L}_2^{nN}).$$

The canonical map

$$H^0(X, \mathcal{F} \otimes \mathcal{L}_1^n) \otimes S \longrightarrow H^0(X, \mathcal{F} \otimes \mathcal{L}_1^n \otimes \mathcal{M}^n)$$

is injective for the nonzero section $S \in H^0(X, \mathcal{M}^n)$ because $\mathcal{M}^n$ is locally free. But $\dim H^0(X, \mathcal{M}) \geq 1$, so $\dim H^0(X, \mathcal{M}^n) \geq 1$. This implies that

$$\dim H^0(X, \mathcal{F} \otimes \mathcal{L}_1^n) \leq \dim H^0(X, \mathcal{F} \otimes \mathcal{L}_2^{nN})$$

or

$$h_1(n) \leq h_2(nN) \quad \text{for any } n \geq n_0.$$

By symmetry, $h_2(n) \leq h_1(nN')$, so $\deg h_1 = \deg h_2$. The theorem is proved. $\qquad \square$

Now let us give another, often useful, description of the dimension. Recall that $f \in R$ is said to be an *essential zero divisor* in the graded R-module M, if the kernel of the multiplication by f ($f: M \xrightarrow{f} M$) has infinitely many nonzero homogeneous components, i.e., is non-trivial in the category GM_R.

2.12.3 M-sequences

Let M be a graded R-module. A finite set of homogeneous and non-invertible elements $\{f_0, \ldots, f_d\}$ in R such that f_i is not an essential zero divisor in $M/(f_0, \ldots, f_{i-1})$ for all $i \geq 0$ is called an *M-sequence*.

The number d is the *length* of the M-sequence $\{f_0, \ldots, f_d\}$. A given M-sequence is said to be *maximal* if it is impossible to add to it any element of R and still get an M-sequence.

The symbol $d(M)$ will denote the *length of the shortest of the maximal M-sequences*.

2.12.4 Theorem $d(M) = \dim \widetilde{M}$.

We will need the following result.

2.12.4a Lemma *If* $\dim \widetilde{M} \geq 0$, *then there exists an element* $f \in R$ *such that* $\{f\}$ *is an M-sequence.*

We will prove this lemma later; now let us use it to prove Theorem 2.12.4.

2.12.5 Proof of Theorem 2.12.4

We use induction on $d(M)$, starting with the case $d(M) = -1$, in which there are no M-sequences.

For $d(M) = -1$, Lemma 2.12.4a immediately implies that $\dim \widetilde{M} = -1$.

Suppose the theorem is proved for all M with $d(M) \leq d-1$ and let $d(M) = d$. Let $\{f_0, \ldots, f_d\}$ be the shortest maximal M-sequence. Then $d(M/f_0 M) = d-1$, and by the induction hypothesis $\dim \widetilde{(M/f_0 M)} = d-1$. Consider the exact sequence

$$0 \longrightarrow N \longrightarrow M \stackrel{f_0}{\longrightarrow} M(k) \longrightarrow M/f_0 M \longrightarrow 0,$$

where $k = \deg f_0$ and $M \stackrel{f_0}{\longrightarrow} M(k)$ is the grading-preserving homomorphism of multiplication by f_0. As $f_0 \in R$ is not an essential zero divisor, we have $N_n = 0$ for $n \geq n_0$ and $\dim M(k)_n - \dim M_n = \dim(M/f_0 M)_n$. Therefore $\dim \widetilde{M} = (d-1)+1 = d$, as needed. $\qquad\qquad\qquad\qquad\qquad\qquad\qquad\qquad\qquad\qquad\qquad\qquad\qquad\square$

2.12.6 Proof of Lemma 2.12.4a

In R, we have to find an element that is not an essential zero divisor in M.

First of all, let us show that there exists a sequence of graded modules M_i such that

$$0 = M_0 \subset M_1 \subset \cdots \subset M_r = M, \quad M_i/M_{i+1} \simeq (R/\mathfrak{p}_i)_{n_i},$$

where the $\mathfrak{p}_i$ are prime graded ideals.

Since M is a Noetherian module, it suffices to find a non-trivial graded submodule $M_1 \subset M$ such that $M_1 \simeq (R/\mathfrak{p})_n$, where $\mathfrak{p}$ is a prime graded ideal.

Let S be the set of graded ideals $\mathfrak{p}_m$ in R for each of which there exists a homogeneous element $m \in M$ whose annihilator is $\mathfrak{p}_m$. Since R is Noetherian, there is a maximal element in S; denote it $\mathfrak{p}$. Clearly, $R_m \simeq (R/\mathfrak{p})_n$, where $\mathfrak{p} = \mathrm{Ann}\, m$, $n = \deg m$. Let us prove that $\mathfrak{p}$ is prime. Indeed, let $ab \in \mathfrak{p}$, $a \notin \mathfrak{p}$. Then the inclusion

Ann(bm) $\supset$ (a, Ann m) is strict by maximality, so $bm = 0$ and $b \in$ Ann m. Hence, $\mathfrak{p}$ is prime.

(Observe that the same argument without taking grading into account proves a similar result for non-graded Noetherian modules.)

We have thus constructed the requisite sequence of modules. Now we use it to find in R an inessential zero divisor in M. Let $\mathfrak{q}$ be a maximal graded prime ideal. If $\mathfrak{q} \not\subset \bigcup \mathfrak{p}_i$, then, for an element of R to be found, we can take any element of $\mathfrak{q}$ not lying in $\bigcup \mathfrak{p}_i$. (Since it lies in $\mathfrak{q}$, it is non-invertible.) If, however, $\mathfrak{q} \subset \bigcup \mathfrak{p}_i$ for any maximal ideal $\mathfrak{q}$, then $\mathfrak{q} \subset \mathfrak{p}_j$ for some j because the ideals $\mathfrak{p}_i$ are prime, so the ideals $\mathfrak{p}_i$ exhaust the set of maximal ideals. Since there are finitely many of them, $\dim R = 0$.

In this case $R \approx \Gamma(X, \mathcal{O}_X)[T]$ is the polynomial ring in one indeterminate T (here "$\approx$" means *isomorphism up to a finite number of homogeneous components*) because $\dim R = \text{const}$ for $n \geq n_0$. Therefore, multiplication by T has no kernel in all sufficiently large dimensions, and T is the sought-for non-essential zero divisor.

Both Lemma 2.12.4a and Theorem 2.12.4 are completely proved. $\square$

In the course of the proof we have obtained a number of useful statements; we separate them for convenience of references.

2.12.7 Corollary

(1) Ann $M = \bigcup \mathfrak{p}_i$, where $(\mathfrak{p}_i)_{i \in I}$ is a finite family of prime ideals in R.
(2) Let $\mathcal{F}$ be a coherent sheaf on $X = \text{Proj}\, R$. Then there exists a sequence of coherent subsheaves

$$0 = \mathcal{F}_0 \subset \mathcal{F}_1 \subset \mathcal{F}_2 \subset \cdots \subset \mathcal{F},$$

such that $\mathcal{F}_{i+1}/\mathcal{F}_i \simeq \mathcal{O}_X/\mathcal{J}_i(n_i)$, where $\mathcal{J}_i$ are coherent sheaves of prime ideals on X.

2.12.8 Theorem $H^q(X, \mathcal{F}) = 0$ for $q > \dim \mathcal{F}$.

Proof Suppose the theorem be established for some sheaves $\mathcal{F}'$ and $\mathcal{F}''$ entering the exact sequence

$$0 \longrightarrow \mathcal{F}' \longrightarrow \mathcal{F} \longrightarrow \mathcal{F}'' \longrightarrow 0.$$

Then

$$\chi(\mathcal{F}(n)) = \chi(\mathcal{F}'(n)) + \chi(\mathcal{F}''(n)),$$

and hence $\dim \mathcal{F} = \max(\dim \mathcal{F}', \dim \mathcal{F}'')$. The exact cohomology sequence easily implies the validity of the statement of the theorem for $\mathcal{F}$: it suffices to consider the terms

$$\cdots \longrightarrow H^q(X, \mathcal{F}') \longrightarrow H^q(X, \mathcal{F}) \longrightarrow H^q(X, \mathcal{F}'') \longrightarrow 0, \quad \text{where } q > \dim \mathcal{F}.$$

Set $\mathcal{F} := \mathcal{O}_X/\mathcal{J}(n)$, where $\mathcal{J} \subset \mathcal{O}_X$ is a coherent sheaf of ideals. The sheaf $\mathcal{J}$ determines a closed subscheme $Y \subset X$, where $\mathcal{O}_Y = \mathcal{O}_X/\mathcal{J}$, which can be considered as the projective spectrum of the graded ring $\bigoplus_{n=0}^{\infty} \Gamma(X, \mathcal{O}_X/\mathcal{J}(n))$.

Therefore, and thanks to item (2) of Corollary 2.12.7, it suffices to verify the statement for $\mathcal{O}_Y(n) = \mathcal{F}$. Set $J := \operatorname{Im} \Gamma_*(X, \mathcal{J}) \subset R$.

Let $d = \dim Y$. By Theorem 2.12.4 there exists a maximal R/J-sequence $f_0, \ldots, f_d$, where $f_i \in R$, such that $\dim_k R/(f_0, \ldots, f_d, J) < \infty$. So setting $\bar{f}_i = f_i$ (mod J) we have $(R/J)_n \subset (\bar{f}_0, \ldots, \bar{f}_d)$ for $n \geq n_0$. Geometrically, this means that $Y = \bigcup_{i=0}^{d} D_+(\bar{f}_i)$. Computing the Čech cohomology immediately yields the statement of the theorem. $\qquad\square$

2.12.8a Remark Since $\dim \mathcal{F} \leq \dim X$ for any $\mathcal{F}$ (as follows from the proof of Theorem 2.12.8), one concludes that, in particular, $H^q(X, \mathcal{F}) = 0$ for any $q > \dim X$.

2.13 The Grothendieck Group: First Notions

2.13.1 The Riemann-Roch Problem

The classical "Riemann-Roch problem" is: **compute** $\dim H^0(X, \mathcal{F})$, **where X is a projective scheme over a field, and $\mathcal{F}$ is a coherent sheaf on X.** The main qualitative information about this function is the statement

$$\dim H^0(X, \mathcal{F}(n)) \text{ is a polynomial, } \chi, \text{ for } n \geq n_0(\mathcal{F}).$$

In practice, therefore, the Riemann-Roch problem is usually subdivided into two questions, solved by distinct approaches:

(a) Describe the coefficients of the Hilbert polynomial of $\mathcal{F}$ in "geometric" terms. $\qquad$ (2.18a)

(b) Find a "good" estimate of the number $n_0(\mathcal{F})$. $\qquad$ (2.18b)

As an example of answer to question (2.18a) we offer the description of the degree of a given projective scheme $X \subset \mathbb{P}^r$ by means of Bézout's theorem: this degree is equal to the number of intersection points of X with a "sufficiently general" linear submanifold of $\mathbb{P}^r_k$ of complementary dimension. The general answer is given by the Riemann–Roch–Hirzebruch–Grothendieck–... theorem.

In these lecture notes, we will not touch upon question (2.18b).

The rest of the chapter is devoted to the study of the characteristic $\chi(\mathcal{F})$. Lemma 2.12.1 describes its main property. Axiomizing this property, we introduce the following definition.

2.13.2 Additive Functions on Abelian Categories

Let X be a scheme, Coh_X the category of coherent sheaves on it. Let G be an Abelian group; a map $\psi \colon \mathsf{Coh}_X \longrightarrow G$ is said to be an *additive function on* Coh_X (or any other Abelian category) *with values in G* if, for any exact sequence

$$0 \longrightarrow \mathcal{F}_1 \longrightarrow \mathcal{F} \longrightarrow \mathcal{F}_2 \longrightarrow 0$$

of sheaves in the category Coh_X, it holds that

$$\psi(\mathcal{F}) = \psi(\mathcal{F}_1) + \psi(\mathcal{F}_2).$$

Additive functions enjoy the following properties.

2.13.2a Lemma *Let ψ be an additive function on* Coh_X, *and $\mathcal{F}_i$ objects of* Coh_X.

(a) *For any exact sequence*

$$0 \longrightarrow \mathcal{F}_1 \longrightarrow \cdots \longrightarrow \mathcal{F}_s \longrightarrow 0,$$

we have

$$\sum_{i=1}^{s} (-1)^i \psi(\mathcal{F}_i) = 0.$$

(b) *Let $0 = \mathcal{F}_0 \subset \mathcal{F}_1 \subset \cdots \subset \mathcal{F}_s = \mathcal{F}$. Then*

$$\psi(\mathcal{F}) = \sum_{i \geq 1} \psi(\mathcal{F}_i / \mathcal{F}_{i-1}).$$

2.13.2b Exercise Prove Lemma 2.13.2a.

It is not difficult to see that even for the simplest schemes X (for example, for $X = \operatorname{Spec} K$), there are plenty of different additive functions on Coh_X. Nevertheless, the entire set of such functions can be easily surveyed thanks to the existence of a "universal" additive function $\mathrm{k} \colon \mathsf{Coh}_X \longrightarrow K(X)$ with values in a universal group $K(X)$.

Let $\mathbb{Z}[\mathsf{Coh}_X]$ be the free Abelian group generated by the symbols $[\mathcal{F}]$ corresponding to the classes of isomorphic coherent sheaves on X. Let $J \subset \mathbb{Z}[\mathsf{Coh}_X]$ be the subgroup generated by the elements

$$[\mathcal{F}] - [\mathcal{F}_1] - [\mathcal{F}_2],$$

one such element for each exact sequence

$$0 \longrightarrow \mathcal{F}_1 \longrightarrow \mathcal{F} \longrightarrow \mathcal{F}_2 \longrightarrow 0.$$

2.13.3 The Grothendieck Group $K(X)$

The group $K(X) := \mathbb{Z}[\mathsf{Coh}_X]/J$ is called the *Grothendieck group* (of the category Coh_X or the scheme X).

2.13.3a Proposition *The function*

$$k: \mathsf{Coh}_X \longrightarrow K(X), \tag{2.19}$$

given by $k(\mathcal{F}) = [\mathcal{F}]$ (mod J) *is additive; the image* $k(\mathsf{Coh}_X)$ *generates the group* $K(X)$, *and for any additive function* $\psi: \mathsf{Coh}_X \longrightarrow G$, *there exists a uniquely determined homomorphism* $\varphi: K(X) \longrightarrow G$ *such that* $\psi = \varphi \circ k$.

Proof is trivial.   □

In terms of the group $K(X)$ so defined, the *Riemann-Roch problem* is

$$\text{describe the function } \chi: K(X) \longrightarrow \mathbb{Z}.$$

The advantage of such a reformulation of Riemann-Roch problem is as follows. As we will see shortly, the group $K(X)$ is endowed with a rich collection of additional structures. Sometimes $K(X)$ can be explicitly and completely computed (for example, for $X = \mathbb{P}_K^r$), and then we can describe χ using this information. In the general case, sufficient information on $K(X)$ is known to geometrically interpret $\chi(\mathcal{F})$.

Lemma 2.13.2a gives an approach to computing $K(X)$ which sometimes allows one to exhibit a smaller set of generators of $K(X)$ than the set $\{k(\mathcal{F})$ for *all* coherent sheaves $\mathcal{F}\}$. Let us illustrate this by examples.

2.13.4 Examples

(1) Let A be any Noetherian ring, $X = \mathrm{Spec}\, A$. In Sect. 2.12.6 we established that any Noetherian A-module has a composition series with factors isomorphic to $A/\mathfrak{p}$, where $\mathfrak{p} \subset A$ is a prime ideal. The generators of $K(X)$ are, therefore, the elements $k(A/\mathfrak{p})$.

To describe all relations is a bit more difficult. We confine ourselves to the case where A is an *Artinian* ring. In this case, the Jordan-Hölder theorem can be interpreted as the computation of $K(X)$: *the map*

$$j: K(X) \longrightarrow \mathbb{Z}[X],$$

where $\mathbb{Z}[X]$ *is the free Abelian group generated by the points of the scheme X, and*

$$j \colon k(\mathcal{F}) \longmapsto \sum_{x \in X} \mathrm{length}_{\mathcal{O}_x} \mathcal{F}_x$$

is a group isomorphism.

(2) Let A be a principal ideal ring, $X = \operatorname{Spec} A$. Then any Noetherian A-module M has a *free projective resolution* of length 1:

$$0 \longrightarrow A^r \longrightarrow A^s \longrightarrow M \longrightarrow 0.$$

This immediately implies that the group $K(X)$ is cyclic and generated by the class of the ring A. The order of this class is equal to infinity, which is easy to see passing to linear spaces $M \otimes_K A$ over the ring of fractions K of A; hence,

$$K(X) \simeq \mathbb{Z}.$$

More generally, *the same is true for any affine scheme* $\operatorname{Spec} A$ *provided any Noetherian A-module has a free resolution of finite length.*

This condition is still too strong to lead to interesting notions; however, a slight weakening of it defines a very important class of schemes.

2.13.5 Smooth Schemes

Let X be a Noetherian scheme, $\mathcal{F}$ a coherent sheaf on it. Suppose that for any point $x \in X$ there exists an open neighborhood $U \ni x$ such that the sheaf $\mathcal{F}|_U$ has in U a finite resolution consisting of "free" sheaves $\mathcal{O}_X^r|_U$. Then the scheme X is said to be *smooth*.

In the next section, two types of schemes will be shown to be smooth: projective spaces over fields, and spectra of local rings whose maximal ideals are generated by regular sequences of elements.

Smoothness of projective spaces immediately follows from the following classical theorem.

2.13.6 Theorem (Hilbert's Syzygy[11] Theorem) *Let* $R = K[T_0, \ldots, T_r]$. *Any graded R-module has a graded projective free resolution of length* $\leq r + 1$.

---

[11]In broadest terms, *syzygy* is a kind of unity, especially through coordination or alignment, most commonly used in the astronomy and astrology. In mathematics, a syzygy is a relation between generators of a module M. The set of all such relations is called the *first syzygy module of M*. A relation between generators of the first syzygy module is called a *second syzygy* of M, and the set of all such relations is called the *second syzygy module of M*, and so on.

The proof will be given at the end of the chapter; now we use this theorem to compute $K(\mathbb{P}_K^r)$.

2.13.7 Theorem *Let* k *be the function defined in* (2.19).
The map $x^i \longmapsto k(\mathcal{O}(i))$ *defines an isomorphism of Abelian groups*

$$\mathbb{Z}[x]/((x-1)^{r+1}) \longrightarrow K(\mathbb{P}_K^r).$$

In particular, the group $K(\mathbb{P}_K^r)$ *is free of rank* $r + 1$.

Proof Translating the first statement of Theorem 2.13.6 into the language of sheaves by means of Theorem 2.11.6, we find that any coherent sheaf $\mathcal{F}$ on $\mathbb{P}_K^r$ possesses a resolution whose terms are direct sums of sheaves $\mathcal{O}(n)$, where $n \in \mathbb{Z}$. Lemma 2.13.2a shows that the elements $k(\mathcal{O}(n))$ generate the group $K(\mathbb{P}_K^r)$.

Obviously, these generators are not independent. At least one relation is obtained if, using Proposition 2.9.4, we consider the Koszul complex $K(T, R) := K(T_0, \dots, T_R; R)$, which is a resolution of the R-module $M = R/(T_0, \dots, T_n)$, where

$$K_i(T, R) := R^{\binom{r+1}{i}} = \wedge^i(Re_0 + \dots + Re_r) :$$
$$\dots \longrightarrow R^{\binom{r+1}{i}} \longrightarrow \dots \longrightarrow R^{\binom{r+1}{2}} \longrightarrow R^{r+1} \longrightarrow R \longrightarrow M \longrightarrow 0.$$

We can regard this resolution as an exact sequence of graded modules if we regard the elements $e_{i_1} \wedge \dots \wedge e_{i_k}$ as homogeneous of degree k. Applying to this resolution the *sheafification functor*, we get the exact sequence

$$0 \longrightarrow \dots \longrightarrow \mathcal{O}(-i)^{\binom{r+1}{i}} \longrightarrow \dots \longrightarrow \mathcal{O}(-2)^{\binom{r+1}{2}} \longrightarrow \mathcal{O}(-1)^{r+1} \longrightarrow \mathcal{O}_{\mathbb{P}^r} \longrightarrow 0$$

(take into account that the R-module M is *TN*-isomorphic to 0, i.e., $\widetilde{M} = 0$). This exact sequence can be tensored by $\mathcal{O}(n)$ for any $n \in \mathbb{Z}$ without violating the exactness. Therefore, the following relations hold in $K(\mathbb{P}_K^r)$:

$$\sum_{i=0}^{r+1} \binom{r+1}{i} k(\mathcal{O}(n-i)) = 0.$$

This clearly implies that the kernel of the homomorphisms of additive groups

$$\mathbb{Z}[x, x^{-1}] \longrightarrow K(X), \quad x^i \longmapsto k(\mathcal{O}(i)) \text{ for } i \in \mathbb{Z},$$

contains the ideal generated by the polynomial $(x^{-1} - 1)^{r+1}$ or, equivalently, by $(x-1)^{r+1}$.

Since we have already established that this homomorphism is an epimorphism, to complete the proof of Theorem, it suffices to verify that the elements $k(\mathcal{O})$, $k(\mathcal{O}(1))$, $\dots$, $k(\mathcal{O}(r))$ are linearly independent over $\mathbb{Z}$.

Obviously, the functions $\chi_n(\mathcal{F}) := \chi(\mathcal{F}(n))$ are additive on $K(\mathbb{P}^r_K)$ for any $n \in \mathbb{Z}$. Therefore, the assumption that there exists a non-trivial linear dependence

$$\sum_{i=0}^{r} a_i k(\mathcal{O}(i)) = 0, \text{ where } a_i \in \mathbb{Z},$$

would imply that

$$\sum_{i=0}^{r} a_i \chi_n(\mathcal{O}(i)) = \sum_{i=0}^{r} a_i \binom{n+i+r}{r} = 0,$$

which is only possible if $a_i = 0$ for all i because, as an easy verification shows (say, by induction on r), the polynomials $\binom{n+i+r}{r}$ in n are linearly independent for $i = 0, \dots, r$. $\square$

2.13.8   The Group $K(\mathbb{P}^r_K)$, see Theorem 2.13.7, is Endowed with a Ring Structure

The multiplication in this ring has an invariant meaning: indeed, as is easy to see,

$$k(\mathcal{F})k(\mathcal{O}_{\mathbb{P}^r}(i)) = k(\mathcal{F}(i)) = k(\mathcal{F} \otimes \mathcal{O}_{\mathbb{P}^r}(i)),$$

so this multiplication corresponds, at least sometimes, to the tensor products of sheaves. There are, however, examples in which $v(\mathcal{F})k(\mathcal{G}) \neq k(\mathcal{F} \otimes \mathcal{G})$, so the general description of the multiplication can not be that simple. This question is studied in detail in the second part of the course, see [Ma3].

Meanwhile, using our description of $K(\mathbb{P}^r_K)$, we give a (rather naive) form of the Riemann-Roch theorem for the projective space.

The idea is to select some simple additive function on $K(\mathbb{P}^r_K)$, and then "propagate" it using ring multiplication.

Every element of $K(\mathbb{P}^r_K)$ can be uniquely represented, thanks to Theorem 2.13.7, as a polynomial $\sum_{i=0}^{r} a_i (l-1)^i$, where $l = k(\mathcal{O}(1))$. Introduce the additive function $\varkappa_r: K(\mathbb{P}^r_K) \longrightarrow \mathbb{Z}$ by setting

$$\varkappa_r \left(\sum_{i=0}^{r} a_i (l-1)^i \right) = a_r.$$

2.13.9 Lemma *For any additive function*

$$\psi: K(\mathbb{P}^r_K) \longrightarrow \mathbb{Z}$$

there exists a unique element $t(\psi) \in K(\mathbb{P}_K^r)$ such that

$$\psi(y) = \varkappa_r(t(\psi)y), \text{ for all } y \in K(\mathbb{P}_K^r).$$

Proof The function ψ is a linear combination of the coefficients a_i appearing in the representation $\sum_{i=0}^r a_i(l-1)^i$, considered as functions on $K(\mathbb{P}_K^r)$; for these coefficients, we have:

$$a_i(y) = \varkappa_r((l-1)^{r-i}y). \qquad \Box$$

2.13.10 Theorem *Let $\chi: K(\mathbb{P}_K^r) \longrightarrow \mathbb{Z}$ be the Euler characteristic. Then*

$$\chi(y) = \varkappa_r(l^r y), \text{ i.e., } t(\chi) = l^r. \tag{2.20}$$

Proof It suffices to verify the coincidence of the left-hand side of (2.20) and the right-hand side for the elements $y = l^i$, where $i = 0, 1, \ldots, r$, that form a $\mathbb{Z}$-basis of the group $K(\mathbb{P}_K^r)$. We have:

$$\chi(l^i) = \chi(\mathcal{O}(i)) = \dim H^0(\mathbb{P}_K^r, \mathcal{O}(i)) = \binom{r+i}{i},$$

$$\varkappa_r(l^{r+i}) = \varkappa_r((1+(l-1))^{r+i}) = \binom{r+i}{i}. \qquad \Box$$

2.13.10a Remark The usage of $\varkappa_r$ as a simplest additive function was not yet motivated. Moreover, it is clear that to apply Theorem 2.13.10 is not that easy: to compute $\chi(\mathcal{F})$ with its help, we have to first know what is the class of the sheaf $\mathcal{F}$ in $K(\mathbb{P}_K^r)$. The only means known to us at the moment is to consider the resolution of the sheaf $\mathcal{F}$, but this is not "geometric", and also makes our formula redundant: If we know the resolution, we can calculate $\chi(\mathcal{F})$ just by additivity.

Nevertheless, formula (2.20) is very elegant; I consider this as a serious argument in its favor.

2.14 Resolutions and Smoothness

Let us show, first of all, that smoothness of a given scheme X is a property of the entire collection of its local rings $\mathcal{O}_x$.

2.14.1 Theorem *A scheme X is smooth if and only if $\operatorname{Spec} \mathcal{O}_x$ is a smooth scheme for all points $x \in X$.*

Proof Let X be smooth. Consider an arbitrary coherent sheaf on Spec $\mathcal{O}_x$. This sheaf is determined by an $\mathcal{O}_x$-module $\mathcal{F}_x$. Let us show that, in some neighborhood $U \ni x$, there exists a sheaf $\mathcal{F}$ whose fiber at x coincides with $\mathcal{F}_x$. Indeed, consider the exact sequence

$$\mathcal{O}_x^r \xrightarrow{f} \mathcal{O}_x^s \longrightarrow \mathcal{F}_x \longrightarrow 0,$$

where the homomorphism f is defined by an $r \times s$ matrix whose entries are germs of sections of the structure sheaf of X at point x. There exists an affine neighborhood Spec A of x to which the entries of this matrix can be continued. Let

$$f: A^r \longrightarrow A^s$$

be the homomorphism determined by this continuation. Set $M := \operatorname{Coker} f$ and let $\mathcal{F}$ be the sheaf $\widetilde{M}$ considered on Spec A. Its stalk at the point x is isomorphic to $\mathcal{F}_x$. Since the scheme X is smooth, we may assume that $U = \operatorname{Spec} A$ is so small that in it there is a finite resolution

$$0 \longrightarrow \mathcal{O}_x^{r_n}|_U \longrightarrow \cdots \longrightarrow \mathcal{O}_x^{r_0}|_U \longrightarrow \mathcal{F} \longrightarrow 0.$$

Passing to stalks at the point x, we obtain a finite resolution of the stalk $\mathcal{F}_x$ on Spec $\mathcal{O}_x$; this proves that Spec $\mathcal{O}_x$ is smooth.

Conversely, let Spec $\mathcal{O}_x$ be a smooth scheme, $\mathcal{F}$ a coherent sheaf on X. Since the only neighborhood of a closed point x in Spec $\mathcal{O}_x$ is the whole spectrum, it follows that there exists a resolution of the stalk

$$0 \longrightarrow \mathcal{O}_x^{r_n} \longrightarrow \cdots \longrightarrow \mathcal{O}_x^{r_0} \longrightarrow \mathcal{F}_x \longrightarrow 0.$$

An argument analogous to the one above allows us to continue this sequence to an open neighborhood U of the point x:

$$0 \longrightarrow \mathcal{O}_x^{r_n}|_U \longrightarrow \cdots \longrightarrow \mathcal{O}_x^{r_0}|_U \longrightarrow \mathcal{F}|_U \longrightarrow 0.$$

Since this sequence is exact at the point x, it follows that it remains exact in a (perhaps, smaller than U) neighborhood of x, as needed. $\square$

This result indicates the importance of studying smooth spectra of local rings. To this end, we will need several elementary facts of homological algebra.

The next result allows one to examine the family of resolutions of a given A-module M.

2.14.2 Lemma *Let A be a ring, M an A-module, and P and P' either projective or free A-modules. Let*

$$0 \longrightarrow S \longrightarrow P \xrightarrow{f} M \longrightarrow 0$$

and

$$0 \longrightarrow S' \longrightarrow P' \xrightarrow{\ f\ } M \longrightarrow 0$$

be two exact sequences. Then $S \oplus P' \simeq S' \oplus P$.

Proof Consider the commutative diagram

where $T = \mathrm{Ker}(f + f')$. Clearly, g is an epimorphism; as P' is projective, there exists a "section" $s: P' \longrightarrow T$ (i.e., a homomorphism such that $g \circ s = 1_T$) and $T \simeq \mathrm{Ker}\, g \oplus P'$. But

$$\mathrm{Ker}\, g = \{(p, 0) \mid f(p) = 0\} \simeq S,$$

so $T \simeq S \oplus P'$. By symmetry, $T \simeq S' \oplus P$, proving the lemma.   □

This result is a reason to introduce the following definition.

2.14.3 Projective or Free Equivalences of Modules

Two A-modules S and S' are said to be *projectively* (respectively *freely*) *equivalent*, if there exist projective (resp., free) A-modules P and P' such that

$$S \oplus P' \simeq S' \oplus P.$$

We can now sharpen Lemma 2.14.2 as follows.

2.14.3a Lemma *Let the A-modules M and M' be projectively (resp., freely) equivalent and let exact sequences*

$$0 \longrightarrow S \longrightarrow P \longrightarrow M \longrightarrow 0,$$
$$0 \longrightarrow S' \longrightarrow P' \longrightarrow M' \longrightarrow 0,$$

be given, where P and P' are projective (resp., free) modules. Then S and S' are projectively (resp., freely) equivalent.

Proof Let $M \oplus Q \simeq M' \oplus Q'$, where Q and Q' are projective (resp., free). Then the following sequences are exact (with obvious homomorphisms)

$$0 \longrightarrow S \longrightarrow P \oplus Q \longrightarrow M \oplus Q \longrightarrow 0,$$

$$0 \longrightarrow S' \longrightarrow P' \oplus Q' \longrightarrow M' \oplus Q' \longrightarrow 0,$$

implying that $S \oplus P' \oplus Q' \simeq S \oplus P \oplus Q$ by Lemma 2.14.2.   □

2.14.3b Corollary *Let there be given the initial segment of a projective (resp., free) resolution of an A-module M,*

$$P_n \xrightarrow{d_n} P_{n-1} \longrightarrow \cdots \longrightarrow P_0 \longrightarrow M.$$

Then the projective (resp., free) equivalence class of the module $\operatorname{Ker} d_n$ *is uniquely determined and does not depend on the choice of the resolution. In particular, it there exists a projective resolution of length n of the module M, then* $\operatorname{Ker} d_n$ *is projective.*

2.14.4   Minimal Resolutions

For modules over local rings, we construct a special type of resolutions, so-called *minimal resolutions*; Corollary 2.14.3b enables us to derive from these minimal resolutions information about arbitrary resolutions.

Let A be a local Noetherian ring, $\mathfrak{m}$ its maximal ideal, M an A-module of finite type, F a free A-module.

An epimorphism $F \longrightarrow M \longrightarrow 0$ is said to be *minimal*, if it induces an isomorphism $F/\mathfrak{m}F \longrightarrow M/\mathfrak{m}M$.

A minimal epimorphism $F \longrightarrow M \longrightarrow 0$ is uniquely determined in the following sense. If $F' \longrightarrow M \longrightarrow 0$ is another minimal epimorphism, then there exists a commutative diagram

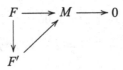

in which the vertical arrow is an isomorphism. Indeed, $\operatorname{rk} F = \operatorname{rk} F'$ by Corollary 1.10.9b and the preimages of any basis of $M/\mathfrak{m}M$ in F and F' are bases in F and F', respectively.

Iterating minimal epimorphisms, we arrive at the notion of *minimal resolution* (here A is supposed to be *Noetherian*): *the resolution*

$$\cdots \longrightarrow F_n \xrightarrow{d_n} F_{n-1} \longrightarrow \cdots \longrightarrow F_0 \longrightarrow M \longrightarrow 0;$$

is minimal if $F_n \xrightarrow{d_n} d_n(F_n)$ *is a minimal epimorphism for all* n.

Claim A resolution is minimal if and only if

$$d_n(F_n) \subset \mathfrak{m}F_{n-1} \quad \text{for all} \quad n \geq 1.$$

2.14.4a Exercise Prove this claim.

2.14.5 Examples

(1) *A minimal resolution.* Let $f_1, \dots, f_r \subset \mathfrak{m}$ be a regular sequence of elements. Then the Koszul complex $K_0(f, A)$ is a minimal resolution of the A-module $A/(f_1, \dots, f_r)$.

(2) *An infinite minimal resolution.* Consider the local ring $A = k + t^2 k[[t]]$, and let $M = \mathfrak{m} = t^2 k[[t]] \subset A$. Then $M = At^2 + At^3$, and $\mathfrak{m}^2 = \mathfrak{m}M = t^4 k[[t]]$, and hence, $\dim_k \mathfrak{m}/\mathfrak{m}^2 = 2$. The initial segment of the minimal resolution is as follows:

$$Af_1 \oplus Af_2 \xrightarrow{d_1} Ae_1 \oplus Ae_2 \xrightarrow{\varepsilon} M \longrightarrow 0,$$

$$\varepsilon(e_i) = t^{i+1},$$

$$d_1(f_1) = t^3 e_1 - t^2 e_2,$$

$$d_1(f_2) = t^4 e_1 - t^3 e_2.$$

It is easy to see that $\operatorname{Ker} d_1 \simeq M$; so in the extension of the resolution this segment will be periodically repeated.

Let us prove now the following theorem on smooth local rings.

2.14.6 Theorem *Let A be a local Noetherian ring whose maximal ideal $\mathfrak{m}$ is generated by a regular sequence $x_1, \dots, x_d$. Then any Noetherian A-module M has a free resolution of length $\leq d$.*

2.14.6a Lemma *Under assumptions of Theorem 2.14.6, construct a minimal resolution $(F_n, d_n)_{n \in \mathbb{Z}_+}$ of the module M and set*

$$S_A^n(M) = \operatorname{Ker} d_n.$$

Then for any $x \in \mathfrak{m}$ which is not a zero divisor in A or in M, we have:

$$S_{A/xA}^n(M/xM) \simeq S^n(M)/xS^n(M).$$

Proof We can confine ourselves to the case where $n = 1$. Consider the commutative diagram

$$
\begin{array}{ccccc}
F & \xrightarrow{\varphi} & M & \longrightarrow & 0 \\
\downarrow & & \downarrow & & \\
F/xF & \xrightarrow{\psi} & M/xM & \longrightarrow & 0
\end{array}
$$

where φ is the minimal epimorphism. The definition easily implies that ψ is also minimal (as an epimorphism of A/xA-modules). Therefore, $S_A^1(M) = \operatorname{Ker} \varphi$, and $S_{A/xA}^1(M/xM) = \operatorname{Ker} \psi$.

There exists a unique homomorphism θ for which the diagram

$$
\begin{array}{ccccccc}
S_A^1(M) & \longrightarrow & F & \xrightarrow{\varphi} & M & \longrightarrow & 0 \\
\downarrow{\scriptstyle\theta} & & \downarrow & & \downarrow & & \\
S_{A/xA}^1(M/xM) & \longrightarrow & F/xF & \xrightarrow{\psi} & M/xM & \longrightarrow & 0
\end{array}
$$

is commutative.

Let us show that θ is an epimorphism. Take $f + xF \in \operatorname{Ker} \psi$. Then $\varphi(f) \in xM$, and hence

$$
\varphi(f - xf_0) = 0 \Longrightarrow f \in xf_0 + S^1(M) \Longrightarrow f + xf \in \theta(S^1(M)).
$$

It remains to verify that $\operatorname{Ker} \theta = xS^1(M)$. Indeed, $\operatorname{Ker} \theta = S^1(M) \cap xF$; but if $\varphi(xf) = 0$, then $\varphi(f) = 0$ because x is not a zero divisor in M. $\square$

2.14.7 Proof of Theorem 2.14.6

Let $\mathfrak{m} = Ax_1 + \cdots + Ax_d$, where $(x_1, \ldots, x_d)$ is a regular sequence. Let us show by induction on d that

$$
S_A^{d+1}(M) = 0 \quad \text{for all } M. \tag{2.21}
$$

If $d = 0$, then A is a field and all is evident.

Suppose the statement (2.21) is true for any regular sequence of $d - 1$ elements. We have

$$
S_A^{d+1}(M) = S_A^d(M'), \quad \text{where } M' = S^1(M).
$$

Since M' is a submodule of a free module, it follows that x_1 is not a zero divisor in M'. By the induction hypothesis and thanks to Lemma 2.14.6, we have

$$0 = S^d_{A/x_1 A}(M'/x_1 M') = S^d_A(M')/x_1 S^d_A(M').$$

The Nakayama lemma implies that

$$S^d_A(M') = 0. \qquad \square$$

2.14.7a Remark The number d is an invariant of the ring A: Indeed, it can be defined as the length of the minimal resolution of the A-module $A/\mathfrak{m}$ (e.g., the *Koszul complex* is such a resolution).

Proof (Sketch) of Hilbert's Syzygy Theorem 2.13.6 Consider a graded ring R instead of the local ring A, take the ideal $R_+ = \bigoplus_{i \geqslant 1} R_i$ instead of the maximal ideal $\mathfrak{m}$, and assume that "module" means "a graded R-module". Then Nakayama's lemma 1.10.9, Corollary 1.10.9b, and also the two notions—of a minimal epimorphism and a minimal resolution, and Theorem 2.14.6 are applicable to the new situation. All arguments can be repeated literally, except for the formulation and proof of Nakayama's lemma: we have to replace "an ideal not equal to A" with "an ideal contained in R_+" and replace the argument with inversion of $1 - f_1$ by the observation that multiplication by an element of R_+ raises the number of the first non-zero component by 1. $\qquad \square$

References

[A] Atiyah, M.F.: *K*-theory. Notes by D.W. Anderson. Benjamin, Inc., New York (1967)

[AM] Atiyah, M.F., Macdonald, I.: Introduction to Commutative Algebra, ix+128 pp. Addison-Wesley, Reading (1969)

[Bb1] Bourbaki, N.: Commutative Algebra. Springer, Berlin (1989)

[Bb2] Bourbaki, N.: Éléments d'histoire mathématiques. Springer/Hermann, Paris (1969)

[Bb3] Bourbaki, N.: General Topology, Chaps. 1–4, 2nd printing, vii+452 pp. Springer, New York (1998)

[Bla] Blass, A.: Existence of bases implies the axiom of choice. In: Axiomatic Set Theory (Boulder, CO, 1983). Contemporary Mathematics, vol. 31, pp. 31–33. American Mathematical Society, Providence (1984)

[BM] Borisov, D.V., Manin, Yu.I.: Generalized operads and their inner cohomomorphisms. In: Kapranov, M., Kolyada, S., Manin, Yu.I., Moree, P., Potyagailo, L.A. (eds.) Geometry and Dynamics of Groups and Spaces: In Memory of Alexander Reznikov. Progress in Mathematics, vol. 265, pp. 247–308. Birkhäuser, Basel (2008)

[BSh] Borevich, Z.I., Shafarevich, I.R.: Number Theory. Academic, New York, NY (1966) (see revised 3rd edn.: Теория чисел. [The theory of numbers], 504 pp. Nauka, Moscow (1985, in Russian))

[Bu] Buium, A.: Arithmetic Differential Equations. AMS Math Surveys and Monographs, vol. 118 (2005)

[Ch] Chebotarev, N.G.: Theory of Algebraic Functions, 396 pp. OGIZ, Moscow-Leningrad (1948, in Russian)

[Dan] Danilov, V.: Cohomology of algebraic varieties. In: Shafarevich, I.R., Danilov, V., Iskovskikh, V. (eds.) Algebraic Geometry—II: Cohomology of Algebraic Varieties. Algebraic Surfaces. Encyclopaedia of Mathematical Sciences, pp. 5–130. Springer, New York (1989)

[Del] Deligne, P., Etingof, P., Freed, D., Jeffrey, L., Kazhdan, D., Morgan, J., Morrison, D., Witten, E. (eds.) Quantum Fields and Strings: A Course for Mathematicians. Material from the Special Year on Quantum Field Theory held at the Institute for Advanced Study, Princeton, NJ, 1996–1997, vol. 1, xxii+723 pp. American Mathematical Society/Institute for Advanced Study (IAS), Providence, RI/Princeton, NJ (1999)

[E1] Eisenbud, D., Harris, J.: The Geometry of Schemes. Graduate Texts in Mathematics, vol. 197, x+294 pp. Springer, New York (2000)

[E2] Eisenbud, D.: Commutative Algebra: with a View Toward Algebraic Geometry. Graduate Texts in Mathematics, vol. 150, xvi+785 pp. Springer, New York (1995)

© The Author(s) 2018

Y.I. Manin, *Introduction to the Theory of Schemes*, Moscow Lectures 1,
https://doi.org/10.1007/978-3-319-74316-5

[E3]   Eisenbud, D., Harris, J.: Schemes. The Language of Modern Algebraic Geometry. The Wadsworth & Brooks/Cole Mathematics Series, xii+157 pp. Wadsworth & Brooks/Cole Advanced Books & Software, Pacific Grove, CA (1992)

[Ef]   Efetov, K.: Supersymmetry in Disorder and Chaos, xiv+441 pp. Cambridge University Press, Cambridge (1997)

[EGA] Grothendieck, A., Éléments de géométrie algébrique. IV. (French) I. Le langage des schémas. Inst. Hautes Études Sci. Publ. Math. no. 4, 228 pp. (1960);
       II. Étude globale élémentaire de quelques classes de morphismes. Inst. Hautes Études Sci. Publ. Math. no. 8, 222 pp. (1961);
       Étude locale des schémas et des morphismes de schémas. III. Inst. Hautes Études Sci. Publ. Math. no. 28, 255 pp. (1966);
       IV. Étude locale des schémas et des morphismes de schémas IV. Inst. Hautes Études Sci. Publ. Math. no. 32, 361 pp. (1967)

All this and many other works by Alexandre Grothendieck—one of the greatest mathematicians of the XX century—is available at:
http://www.grothendieckcircle.org/.

[EM]   Encyclopedia of Mathematics. Kluwer, Alphen aan den Rijn (1987–1992)

[FFG] Fomenko, A., Fuchs, D.: Homotopic Topology. Graduate Texts in Mathematics, vol. 273, vii+627 pp. Springer, New York (2016);
       Fomenko, A., Fuchs, D., Gutenmakher, V.: Homotopic Topology. Academiai Kiadó, Budapest (1987)

[Gab] Gabriel, P.: Categories abeliénnes. Bull. Soc. Math. France **90**, 323–448 (1962)

[GeKr] Gendenshtein, L.E., Krive, I.V.: Supersymmetry in quantum mechanics. Sov. Phys. Usp. **28**(8), 645–666 (1985)

[GH]   Griffiths, P., Harris, J.: Principles of Algebraic Geometry, A. Wiley, New York (1978)

[GK]   Ginzburg, V., Kapranov, M.: Koszul duality for operads. Duke Math. J. **76**(1), 203–272 (1994)

[GM]   Gelfand, S.I., Manin, Yu.I.: Methods of Homological Algebra. Translated from the 1988 Russian original, xviii+372 pp. Springer, New York (1996)

[God] Godeman, R.: Topologie algébrique et théorie des faisceaux, 283 pp. Hermann, Paris (1998)

[GW]   Görtz, U., Wedhorn, T.: Algebraic Geometry I. Schemes. With Examples and Exercises, 615 pp. Springer, Berlin (2010)

[H]   Hartshorne, R.: Algebraic Geometry. Graduate Texts in Mathematics, vol. 52, xiv+496 pp. Springer, New York (1977)

[Hs]   Halmos, P.: Naive Set Theory. Van Nostrand, Princeton (1960)

[J]   Johnstone, P.T.: Topos Theory. London Mathematical Society Monographs, vol. 10, xxiii+367 pp. Academic/Harcourt Brace Jovanovich, London/New York (1977)

[K]   Kelley, J.L.: General Topology. Reprint of the 1955 edition [Van Nostrand, Toronto, ON]. Graduate Texts in Mathematics, vol. 27, xiv+298 pp. Springer, New York (1975)

[KaS] Kashiwara, M., Schapira, P.: Sheaves on Manifolds. With a chapter in French by Christian Houzel. Corrected reprint of the 1990 original. Grundlehren der Mathematischen Wissenschaften [Fundamental Principles of Mathematical Sciences], vol. 292, x+512 pp. Springer, New York (1994)

[Kas] Kassel, C.: Quantum Groups. Graduate Texts in Mathematics, vol. 155, xii+531 pp. Springer, New York (1995)

[Ku]   Kuroda, S.: A generalization of the Shestakov-Umirbaev inequality. J. Math. Soc. Jpn. **60**(2), 495–510 (2008)

[Kz]   Kunz, E.: Introduction to Commutative Algebra and Algebraic Geometry. Translated from the German by Michael Ackerman. With a preface by David Mumford, xiv+238 pp. Birkhäuser/Springer, New York (2013)

[L0]   Leites, D.: Spectra of graded-commutative rings. Usp. Mat. Nauk **29**(3), 209–210 (1974) (in Russian; for details, see Leites, D. (ed.): Seminar on Supermanifolds.

Reports of Department of Mathematics, no. 1–34, 2100 pp. Stockholm University (1986–1990))

[Lan] Lang, S.: Algebraic Number Theory. Graduate Texts in Mathematics, vol. 110, 2nd edn., xiii+357 pp. Springer, New York (2000)

[Lang] Lang, S.: Algebra. Graduate Texts in Mathematics, vol. 211, revised 3rd edn., xvi+914 pp. Springer, New York (2002)

[M] Macdonald, I.G.: Algebraic Geometry. Introduction to Schemes, vii+113 pp. W.A. Benjamin, New York (1968)

[Ma1] Manin, Yu.I.: Lectures on Algebraic Geometry (1966–68). Moscow University Press, Moscow (1968, in Russian)

[Ma2] Manin, Yu.I.: Lectures on Algebraic Geometry. Part 1. Moscow University Press, Moscow (1970, in Russian)

[Ma3] Manin, Yu.I.: Lectures on the K-functor in algebraic geometry. Russ. Math. Surv. **24**(5), 1–89 (1969)

[Ma4] Manin, Yu.I.: A Course in Mathematical Logic. Translated from the Russian by Neal Koblitz. Graduate Texts in Mathematics, vol. 53, xiii+286 pp. Springer, New York (1977)

[MaD] Manin, Yu.I.: New dimensions in geometry. Russ. Math. Surv. **39**(6), 51–83 (1984)

An expanded version appeared in English in: Workshop Bonn 1984 (Bonn, 1984), Lecture Notes in Mathematics, vol. 1111, pp. 59–101. Springer, Berlin (1985), where it is followed by some interesting remarks by M.F. Atiyah: Commentary on the article of Yu. I. Manin: "New dimensions in geometry" [Workshop Bonn 1984 (Bonn, 1984), pp. 59–101. Springer, Berlin (1985); MR 87j:14030]. Workshop Bonn 1984 (Bonn, 1984), pp. 103–109.

[MaN] Manin, Yu.I.: Numbers as functions. In: P–adic Numbers, Ultrametric Analysis and Applications, vol. 5, no. 4, pp. 313–325 (2013). arXiv:1312.5160

[McL] Mac Lane, S.: Categories for the Working Mathematician. Graduate Texts in Mathematics, vol. 5, 2nd edn., xii+314 pp. Springer, New York (1998)

[M1] Mumford, D.: Lectures on Curves on an Algebraic Surface. With a section by G.M. Bergman. Annals of Mathematics Studies, vol. 59, xi+200 pp. Princeton University Press, Princeton, NJ (1966)

[M2] Mumford, D.: Picard groups of moduli problems. In: Arithmetical Algebraic Geometry (Proceedings of a Conference Held at Purdue University, 1963), pp. 33–81. Harper & Row, New York, NY (1965)

[M3] Mumford, D.: The Red Book of Varieties and Schemes. Lecture Notes in Mathematics, vol. 1358, vi+309 pp. Springer, Berlin (1988)

[MaG] Manin, Yu.I.: Gauge Field Theory and Complex Geometry. Translated from the 1984 Russian original by N. Koblitz and J.R. King. Grundlehren der Mathematischen Wissenschaften [Fundamental Principles of Mathematical Sciences], vol. 289, 2nd edn., xii+346 pp. Springer, Berlin (1997)

[MaT] Manin, Yu.I.: Topics in Noncommutative Geometry, viii+144 pp. Princeton University Press, Princeton, NJ (1991)

[Mdim] Manin, Yu.I.: The notion of dimension in geometry and algebra. Bull. Am. Math. Soc. **43**, 139–161 (2006). arXiv:math.AG/0506012

[Mo] Molotkov, V.: Infinite-dimensional and colored supermanifolds. J. Nonlinear Math. Phys. **17**(Suppl. 01), 375–446 (2010)

[MV] Molotkov, V.: Glutoses: A Generalization of Topos Theory. arXiv:math/9912060

[OV] Onishchik, A., Vinberg, E.: Lie Groups and Algebraic Groups. Translated from the Russian, Springer Series in Soviet Mathematics, xx+328 pp. Springer, Berlin (1990)

[Pa] Palamodov, V.: Cogitations over Berezin's integral. In: Dobrushin, R.L., et al. (eds.) Contemporary Mathematical Physics (F.A. Berezin memorial volume). American Mathematical Society Translation Series 2, vol. 175. American Mathematical Society, Providence, RI (1996)

[Pr]   Prasolov, V.: Polynomials. Translated from the 2001 Russian Second edition by Dimitry
          Leites. Algorithms and Computation in Mathematics, vol. 11, xiv+301 pp. Springer,
          Berlin (2004)
[Q]   Quillen, D.: Projective modules over polynomial rings. Invent. Math. **36**, 167–171
          (1976)
[R1]   Rosenberg, A.: Noncommutative Algebraic Geometry and Representations of Quan-
          tized Algebras. Mathematics and Its Applications, vol. 330, xii+315 pp. Kluwer,
          Dordrecht (1995)
[R2]   Rosenberg, A.: Almost quotient categories, sheaves and localization. In: Leites, D. (ed.)
          Seminar on Supermanifolds-25. Reports of Department of Mathematics, Stockholm
          University (1986–1990), no. 1–34, ca 2100 pp. A version; Kontsevich, M., Rosenberg,
          A.: Noncommutative smooth spaces. In: Gelfand, I.M., Retakh, V.S. (eds.) The Gelfand
          Seminars 1996–1999, pp. 85–108, Birkhäuser, Basel (2000); arXiv:math/9812158; see
          also
          http://www.mpim-bonn.mpg.de/preprints?year=&number=&name=rosenberg&title=
[Reid]   Reid, M.: Undergraduate Commutative Algebra. London Mathematical Society Student
          Texts, vol. 29, xiv+153 pp. Cambridge University Press, Cambridge (1995)
[RF]   Rokhlin, V.A., Fuchs, D.B.: Beginner's Course in Topology. Geometric Chapters.
          Translated from the Russian, Universitext, Springer Series in Soviet Mathematics,
          xi+519 pp. Springer, Berlin (1984)
[Ru]   Rudakov, A.: Marked trees and generating functions with odd variables. Normat **47**(2),
          66–73, 95 (1999)
[S1]   Serre, J.-P.: Algèbre locale, multiplicités: Cours au Collège de France, 1957–1958.
          Lecture Notes in Mathematics, vol. 11, x+160 pp. Springer, Berlin (1997); Corr. 3rd
          printing
[S2]   Serre, J.-P.: Algebraic Groups and Class Fields, Corr. 2nd edn. Springer, Berlin (1997)
[SGA4]   Théorie des topos et cohomologie étale des schémas (SGA4). Un séminaire dirigé
          par M. Artin, A. Grothedieck, J.L. Verdier Lecture Notes in Mathematics, vol. 269.
          Springer, Berlin (1972)
[SGA4 1/2] Deligne, P. et al.: Séminaire de Géométrie Algébrique du Bois-Marie SGA 4 1/2.
          Cohomologie étale. Lecture Notes in Mathematics, vol. 569. Springer, Berlin (1977)
[Sh0]   Shafarevich, I.: Basic Algebraic Geometry. 1. Varieties in Projective Space. Translated
          from the 1988 Russian edition and with notes by Miles Reid, xx+303 pp, 2nd edn.
          Springer, Berlin (1994);
          Basic Algebraic Geometry. 2. Schemes and Complex Manifolds. Translated from the
          1988 Russian edition by Miles Reid, xiv+269 pp, 2nd edn. Springer, Berlin (1994)
[Sh1]   Shafarevich, I.R.: Basic Notions of Algebra. Translated from the Russian by M. Reid.
          Reprint of the 1990 translation [Algebra. I, Encyclopaedia of Mathematical Sciences,
          vol. 11, Springer, Berlin, 1990], iv+258 pp. Springer, Berlin (1997)
[Sh2]   Shafarevich, I.R.: Number Theory. I. Fundamental Problems, Ideas and Theories.
          Translation edited by Parshin, A.N., Shafarevich, I.R. Encyclopaedia of Mathematical
          Sciences, vol. 49, iv+303 pp. Springer, Berlin (1995)
[SoS]   Leites, D. (ed.): Seminar on Supermanifolds. Reports of Department of Mathematics,
          no. 1–34, 2100 pp. Stockholm University (1986–1990);
          Leites D. (ed.): Seminar on Supersymmetry (Algebra and Calculus: Main Chapters,
          vol. 1, 410 pp.) (J. Bernstein, D. Leites, V. Molotkov, V. Shander). MCCME, Moscow
          (2011, in Russian; a version in English is available for perusal)
[StE]   Steenrod, N., Eilenberg, S.: Foundations of Algebraic Topology, 15+328 pp. Princeton
          University Press, Princeton (1952) (second printing, 1957)
[Su]   Suslin, A.: Projective modules over polynomial rings are free. Sov. Math. Dokl. **17**(4),
          1160–64 (1976)
          Suslin, A.: Algebraic K-theory of fields. In: Proceedings of the International Congress
          of Mathematicians, vols. 1, 2, pp. 222–244 (Berkeley, CA, 1986). American Mathemat-
          ical Society, Providence, RI (1987)

[vdW] van der Waerden, B.L.: Algebra. Vol. I. Based in part on lectures by E. Artin and
       E. Noether. Translated from the seventh German edition by Fred Blum and John R.
       Schulenberger, xiv+265 pp. Springer, New York (1991);
       Vol. II. Based in part on lectures by E. Artin and E. Noether. Translated from the fifth
       German edition by John R. Schulenberger, xii+284 pp. Springer, New York (1991)
[VSu] Vassershtein, L., Suslin, A.: Serre's problem on projective modules over polynomial
       rings and algebraic K-theory. Math. USSR Izvestiya, ser. mathem. **40**, 993–1054 (1976)
[We]   Weil, A.: Théorie des points proches sur les variétés différentiables. (French) Géométrie
       différentielle. Colloques Internationaux du Centre National de la Recherche Scien-
       tifique, Strasbourg, 1953, pp. 111–117. Centre National de la Recherche Scientifique,
       Paris (1953)
[Wi]   Witten, E.: An interpretation of the classical Yang-Mills theory. Phys. Lett. **78B**, 394–
       398 (1978)
[ZS]   Zariski, O., Samuel, P.: Commutative Algebra, vol. 1. Springer, Berlin (1979)

Index

K(X) = K[T]/(F), see Ring
 coordinate of the variety, 2
$(\mathfrak{a} : \mathfrak{b})$, the quotient of two ideals, 48
$(C_S)^\circ$, 99
A-Algs, the category of A-algebras, 96
A-Mods, the category of (left) A-modules over
 an algebra A, 96
$A[S^{-1}]$, see localization at a multiplicative set,
 36
A_S, see localization of A at a multiplicative set,
 36
A_f, localizations with respect to S_f, 36
$A_\mathfrak{p}$, localizations with respect to $S_\mathfrak{p}$, 36
B-point of A, see point, 8
$B^1((U_i)_{i \in I}, \mathcal{O}_X^\times)$, 148
D(f), see set, big open, 19
$D_+(f)$, see set, big open, 127
$H^1((U_i)_{i \in I}, \mathcal{O}_X^\times)$, 148
$J \subset \mathbb{Z}[\text{Coh}_X]$, 182
K-algebra, 2
K-algebra homomorphism, 2
K(f), complex, 165
$K_0(f)$, the Koszul complex, 160
L-point of the system of equations X, 2
M-sequence, 178
N(x), the norm of the field k(x), 78
P^X, the point functor, 101
$P^X(Y) := \text{Hom}_\mathbf{C}(X, Y)$, 101
P_X, the point functor, 101
$P_X(Y) := \text{Hom}_\mathbf{C}(Y, X)$, the set of Y-points of
 X, 84
$S^{-1}A$, see localization at a multiplicative set,
 36
V(E), 16
V(F), the variety defined by the ideal (F), 2

X(L), the set of L-points of the system
 of equations X, 2
$X_f := \text{Spec } A_f$, 58
X_{red}, see scheme, reduced, 42
$Z^1((U_i)_{i \in I}, \mathcal{O}_X^\times)$, 148
Δ_X, see Diagonal, 56
$\Omega^1_{B/A}$, the module of (relative) differentials,
 odd, 70
$\bigvee_i Y_i$, see Schemes, quasiunion of, 43
$\chi(X)$, characteristic, of projective scheme, 132
$\chi(\mathcal{F})$, Euler characteristic of the sheaf $\mathcal{F}$, 176
$\deg h_R(n)$, dimension of projective scheme,
 131
Bun$_M$, 99
Rings, the category of (commutative)
 rings (with unit) and their
 homomorphisms, 96
Rings$_R = R$-Algs, 99
Sets, the category of sets and their maps, 96
Top, the category of topological spaces and
 their continuous maps, 96
Top$_X$, 109
Vebun$_M$, 99
μ_n, group of nth roots of unity, 88
$\nu(p^a)$, the number of geometric $\mathbb{F}_{p^a}$-points
 of A, 78
Ann(b), annihilator of an ideal, 48
Diff(M, N), see differential operators, 72
Diff$_k(M, N)$, 72
Ann f, annihilator of an element, 44
Covect$_{B/A}$, the module of (relative)
 differentials, 70
Pic X, the Picard group, 147
Prime A if X = Spec A, the set of prime ideals
 associated with X (or A), 46

© The Author(s) 2018
Y.I. Manin, *Introduction to the Theory of Schemes*, Moscow Lectures 1,
https://doi.org/10.1007/978-3-319-74316-5

Prime X, where $X = \operatorname{Spec} A$, the set of prime
        ideals associated with X (or A), 46
$\operatorname{Spec} A$, 8
$\operatorname{Spm} A$, the set of maximal ideals, 40
$\operatorname{supp} Y$, the support of the scheme
        $Y = \operatorname{Spec} A/\mathfrak{a}$, 41
supp, see support of a sheaf, 146
tr deg, transcendence degree, 82
$\operatorname{ht}(x)$, height of the point x, 18
$\underline{X}(Y) := P_X(Y)$, 102
x_r, 186
ζ-function, 79
$h_M(n)$, Hilbert polynomial, 131
$i(\mathcal{F})$, the sheaf $\mathcal{F}$ considered as a presheaf, 140
$k(x)$, field of fractions, 9
$n(p^a)$, the number of closed points $x \in \operatorname{Spec} A$
        such that $N(x) = p^a$, 78
$p_a(X)$ genus, arithmetic, of projective scheme
        X, 132
r_U^V, restriction map, 108
$\check{H}^p(U, \mathcal{F})$, see Čech cohomology group, 154
$\mathbb{G}_a$, group, additive, 85
$\mathbb{G}_m$, group, multiplicative, 85
$\mathcal{O}_X(n)$, 149
$K(X)$, Grothendieck group, 183
k: $\operatorname{Coh}_X \longrightarrow K(X)$, 183
$C^* := \operatorname{Funct}(C^\circ, \operatorname{Sets})$, 100
Čech cocycle, 148

Algebra
    flat, 61
    integral over A, 35
Annihilator, 44, 48
Arrow, 94
Artinian ring, 14

Bézout's theorem, 137, 139
Bialgebra, 92
Bundle
    normal, 73
    vector, 58
        stably free, 75
Bundle induced, 54

Cartesian square, 53
Category, 94
    big, 95
    small, 95
Center
    of a geometric point, 9
Chain of points, 18

Change of base, 54
Characteristic
    of projective scheme, 132
Chevalley's theorem, 34
Coalgebra, 92
Cocycle
    Čech, 148
Codimension, 67
Complex
    Čech, 153
    acyclic, 155
    Koszul, 159
Component
    embedded, 47
    irreducible of a given Noetherian
        topological space, 21
    isolated, 47
Conormal
    of the diagonal, 70
Convention, 12

DCC, the descending chain condition, 14
Decomposition
    primary, incontractible, 45
Degree
    of a homomorphism, 131
    of the projective scheme, 132
    transcendence, 82
Derivative
    arithmetic, 74
Diagonal, 56
Dimension
    of projective scheme, 131
Dimension of the arithmetic ring, 82
Dirichlet series, 79

Element
    integral over A, 35
Embedding
    closed, of a subscheme, 41
Epimorphism
    minimal, 190
Equivalence
    birational, 123
Euler
    product, 79

Family
    normal (of vector spaces), 66
    of vector spaces, 57
Fermat's descent, 124

Fiber
  geometric, 56
  ordinary, 56
Field
  of fractions, 9
  of quotients, 9
Filter
  on a set, 26
Formula
  Lefschetz, 81
Frobenius morphism, 81
Function
  flat, 63
Functor, 99
  representable, 101
  corepresentable, 101
  faithful, 169
  full, 169
  point, 101
  sheafification, 185

Galois group, 86
Genus
  arithmetic, of projective scheme, 132
Germ
  of a section, 112
  of neighborhoods, 39
Grading
  standard of $K[T]$, 125
Group
  of Čech cohomology, 154
  additive, 85
  Galois, 86
  general linear, 85
  Grothendieck, 183
  linear algebraic, 89
  multiplicative, 85
  Picard, 147
Group scheme
  affine, 84

Harnak's theorem, 5
Height of the point, 18
Hilbert polynomial, 131
Hilbert's Nullstellensatz (theorem on zeros),
  11, 52
Hilbert's syzygy theorem, 193
Homotopy
  chain, 155

Ideal
  homogeneous, 125
  primary, 44
  prime, 8
  radical, 16
Induction
  Noetherian, 21
Intersection
  complete, 66
  complete, geometric, 138
Isomorphism
  up to a finite number of homogeneous
    components, 180

Lefschetz formula, 81
Legendre's theorem, 4
Lemma
  Nakayama, 63
  Zorn, 9
Length of a module, 14
Limit
  direct, 111
  inductive, 111
Localization
  at a multiplicative set, 36

Module
  locally free, 59
  co-normal, 68
  conormal, 66
  defining a family of vector spaces, 57
  graded, 130
  length of, 14
  projective, 59
  simple, 15
Modules
  TN-isomorphic, 175
  equivalent
    freely, 189
    projectively, 189
Morphism, 94
  diagonal, 56
  Frobenius, 81

Nakayama's lemma, 63
Newton's formulas, 7
Newton–Girard formulas, 7
Nilradical, 13
Noether's normalization theorem, 31
Noetherian topological space, 21

Object
    final, 90
Open =open set, 108
Operator
    differential, 72
    differential, symbol of, 72
Order
    on the set of closed subschemes, 42

Partition
    of unity, 115
Partition of unity, 25
Point
    generic, 17, 20
    geometric, 8
Polynomial
    Hilbert, 131
Presheaf, 108, 109
    of modules, 140
Problem
    open, 138
    Riemann-Roch, 181
Product
    absolute, of affine schemes, 54
    fiber, 52, 53

Question, 6
Quotient
    of ideals, 48

Radical $\tau(I)$ of the ideal I, 16
Resolution
    minimal, 190
Riemann-Roch problem, 181, 183
Ring
    of constants, 1
    arithmetic, 78
    Artinian, 14
    Boolean, 5
    coordinate of the variety, 2
    geometric, 78
    Noetherian, 7
    of fractions, 36
    Veronese, 130
Ring of finite type, 32

Scheme, 119
    (ir)reducible, 43
    Noetherian, 44
    affine, 28, 119

reduced, 42
    smooth, 184
Schemes
    quasiunion of, 43
Schemes, intersection of, 42
Section
    of the presheaf, 108
Sequence
    regular, 166
Serre's problem, 75
Serre's theorem, 168
Seshadri's theorem, 75
Set
    big open, 19, 127
    cofinal, 150
    constructible, 34
    inductive, 9, 21
    locally closed, 34
    multiplicative, 25, 36
Sheaf, 109
    invertible, 147
    coherent, 143
    conormal, 73
    of finite type, 143
    quasi-coherent, 142
    structure, 116, 118
    very ample, 177
Solution
    of a system of equations, 1, 2
Space
    connected, 23
    cotangent Zariski, 68
    irreducible, 20
    ringed, 118
    tangent Zariski, 68
    topological
        quasi-compact, 25
Spectrum
    maximal, 40
    projective, 125
Square
    Cartesian, 53
Stalk, 111
Structure, 95
Sub-category
    full, 94
Subgroup
    closed, 88
Subscheme
    closed, 41
    closed, embedded regularly, 66
    locally regularly embedded at $y \in Y$, 67
    primary, 44
Support

of a sheaf, 146
of the scheme, 41
System
    of equations, 1
    direct or inductive, 97
    directed, 111
    inductive, 111
    inverse or projective, 97
Systems
    of equations, equivalent, 2

Theorem
    Bézout, 137, 139
    Cartier, 89
    Chevalley, 34
    Harnak, 5
    Hilbert's on syzygy, 184, 193
    Legendre, 4
    Noether, normalization, 31
    on elementary divisors, 77
    Serre, 168
    uniqueness, 45

Topology
    Zariski, 16
Topos, 97
Transcendence degree, 31
Transformation
    monoidal, with center in an ideal, 129

Ultrafilter, 26
Unit, xii

Vector
    normal, 66
Vector bundle, 58
    over a scheme, 59
Veronese ring, 130

Zero divisor
    essential, 178
Zorn's lemma, 9

Printed in the United States
By Bookmasters